M000201754

LOGIC
AND
LOGOS

LOGIC AND LOGOS

ESSAYS ON SCIENCE, RELIGION AND PHILOSOPHY

WILLIAM S. HATCHER

GEORGE RONALD

OXFORD

GEORGE RONALD Publisher
46 High Street, Kidlington, Oxford OX5 2DN, England

© William S. Hatcher 1990
All Rights Reserved

British Library Cataloguing in Publication Data

Hatcher, William S. *1935*–
    Logic and logos : essays on science, religion and
    philosophy.
    1. Philosophy
    I. Title
    100

ISBN 0–85398–298–8

CONTENTS

Preface   vii

1 Platonism and Pragmatism   1
2 Myths, Models, and Mysticism   19
3 From Metaphysics to Logic
   A Modern Formulation of Avicenna's Cosmological Proof
   of God's Existence   60
4 A Logical Solution to the Problem of Evil   81
5 Science and the Bahá'í Faith   94

Works Cited   123
Notes   126

Preface

As Immanuel Kant states emphatically in the opening passage of his *Critique of Pure Reason*, there can be no doubt that all human knowledge arises from experience. But discovering the ultimate (and, as it turns out, unobservable) ground of this experience, and relating properly to that ground, has been no easy task. However, five to six thousand years of more-or-less continuous social and intellectual history have served to give us a fairly coherent general picture of the basic human condition.

On the one hand, there is the enterprise of relating to reality by constructing mental models of it. We 'fill the gaps' in our immediate experience by using our imagination to conceive of what the structure of unobservable reality might be like. We articulate these mental models in the form of theories whose validity is then tested through further experience. This way of relating to reality has been systematized and generalized and constitutes what is now called 'science'.

On the other hand, our recognition that we ourselves have sprung from the unknown and unobservable, and will return to it at the moment of our death, inspires in us an appropriate sense of our limitations – of being encompassed by a reality greater than ourselves. We have an acute sense of the transcendence of ultimate reality, and also of the inadequacy and relativity of our theories. We are therefore impelled towards transcendence – towards transcendent experience and transcendent knowledge. The systematization and generalization of this quest for transcendence is what we call 'religion'.

Human life on earth is a constant tension between the polarities of these two endeavors, and human discourse a continual dialectic between them. We might say that science is based on a 'minimalist' articulation of reality. In science, the universal law of cause and effect, which is embedded in the very structure of things, is modelled by the logical connection between hypothesis and conclusion in our theories. This connection can be laid bare only by systematic application of the law of parsimony, according to which the existence of imagined (unobservable) entities shall be postulated only when strictly necessary to explain a given portion of reality. Such is *logic*, i.e. the *science* of the *word*, a distillation of pure thought from the richness and diversity of human experience.

Religion represents rather a 'maximalist' articulation of reality, an articulation that seeks to capture as much of it as is humanly possible. Religious discourse is thus laden with multiple, deep and subtle meanings. Central to the religious enterprise – our quest for transcendence – is the phenomenon of revelation, in which the Logos or creative attributes of God Himself are made manifest in the person of a specially chosen human vehicle, a Moses or a Jesus, a Buddha or a Muḥammad. Their revelation is the fullest expression or articulation of reality that we humans can experience, and the most direct link possible with the ultimate ground of reality from which our knowledge springs.

The essays in the present collection represent my own attempt to understand and to relate these maximalist and minimalist articulations of reality, and all the gradations between them. In making this attempt, one naturally faces the problem posed by entrenched ideological positions which see the polarities of science and religion as conflicting rather than complementary, as opposites rather than as parts of a whole. As I have elsewhere explained, I believe that these rigid viewpoints derive primarily from the refusal of religion to recognize and accept the validity of scientific method, on one hand, and from the dogmatic materialistic philosophy that has plagued the practice of modern science, on the other.

However, during the some thirty years since I began thinking and writing in this vein, the rigidity of these entrenched positions has abated to a certain extent — perhaps to a considerable extent. The recent prominence of systems theory, with its stress on a holistic approach to the study of complex phenomena (such as the human brain or sophisticated social structures), may not be the ultimate paradigm that some of its more enthusiastic supporters have claimed, but it represents a significant move away from the intellectual aridity of positivism that so pervaded science during the first half of this century.

I certainly do not claim, and hardly dare even to hope, that my own contribution to the discussion of these issues has had any impact whatsoever on the intellectual and spiritual ethos of the twentieth century. Be that as it may, the benefit I have personally derived from the exercise of thinking through these questions is, for me, a sufficient reason for having done so. Indeed, over the years I have felt at times irresistibly impelled to examine and re-examine these questions from both universal and particular perspectives. Some of these efforts have been previously published while others appear here for the first time. The present collection therefore gives a fairly representative sample of my struggle to come to grips with the various issues herein discussed.

I am deeply grateful to George Ronald for the opportunity afforded me to bring these pieces together in a single volume, and for their tolerance (and even encouragement) of the kind of back-and-forth that any serious attempt to articulate difficult and important concepts involves. I also wish to thank again my wife, Judith, who has willingly endured much in both active and passive support of my efforts, and my brother John who has likewise always sought to encourage me. Finally, I wish to express my thanks to the Association for Bahá'í Studies under whose auspices and encouragement several of these essays were originally presented.

William S. Hatcher
Quebec, Canada
20 October 1989

I

Platonism and Pragmatism

The exponential explosion of scientific and technological progress beginning in the middle of the nineteenth century and accelerating into the twentieth has produced a hiatus, a formidable discontinuity in the evolution of philosophical thought. The magnitude of this discontinuity is due not only to the suddenness with which the explosion has occurred, but more fundamentally to the fact that the historically recent success of science does not seem to be the child of any identifiable philosophy, nor was it predicted by any school of philosophical thought. With a certain degree of deliberate over-simplification, we might characterize this philosophical hiatus by saying that, during the millennia preceding modern science, the basic problem of epistemology was taken to be

I would like to thank the members of the Department of Philosophy at the University of Alberta at Edmonton for thoughtful criticism of an early version of this essay, which was also presented as a paper at the 7th annual meeting of the Society for Exact Philosophy held at McGill University in June 1979. I have likewise benefited from critical comments by the participants in a joint colloquium of the Mathematics and Philosophy Departments, S.U.N.Y. at Buffalo, to which the paper was subsequently presented.

The present essay is a revised and expanded version of the original paper and was first presented at the plenary session of the 12th annual conference of the Association for Bahá'í Studies held at Princeton University, October 22–25, 1987. This is the first publication of any version of the paper.

'How is it possible for us to attain knowledge?', whereas the question has now become 'What is it about modern scientific method and practice that has enabled us to attain knowledge?'. In other words, modern post-scientific epistemology has concerned itself with developing a proper and accurate description of the essentials of scientific method and practice. Such a descriptive epistemology is essentially *a posteriori* and pragmatic. In contrast, pre-modern epistemologies tended rather to be *a priori* and speculative.

Nevertheless, we recognize that pre-modern philosophy was successful in raising and treating in one way or another most of the epistemological and ontological questions raised by scientific practice itself. Indeed, this realization on our part heightens the sense of discontinuity between the pre-modern and the modern, for we might have expected that successful science would have appeared as the obvious offspring of some equally successful philosophy, rather than emerging, as it did, by unplanned and unforeseen fits and starts.

This sense of discontinuity in our philosophical tradition is particularly acute with regard to ontological questions, for we have all learned to live with the fact that practising scientists can be equally successful in research while holding vastly different and even contradictory ontological presuppositions as 'lay' philosophers. At the same time, scientific practice can be influenced by the philosophical presuppositions of its practitioners, as philosophers such as Bunge have pointed out.

The discontinuity provoked by the emergence of modern science appears both vertically, in our history, and horizontally, within the present-day philosophical community. For not only are there philosophical differences among scientifically-minded philosophers, there are members of the philosophical community who continue to labor within the pre-modern philosophical framework, even contending in some instances that the 'knowledge' which science has given us is not really knowledge at all, but only a poor substitute for the Absolute. [1]

Working within what I have called the modern philosophical framework has several obvious advantages. We can check many (but not all) of our philosophical speculations against specific instances of scientific practice. Although the relationship between such speculation on the one hand and concrete practice on the other may not always be clear or simple, the value of such an external point of reference is nevertheless very great and gives a certain empirical spirit to our philosophy. Furthermore, the continual refinement of our philosophical endeavors, coupled with the continued progress of science itself, serves to sharpen the frontier between arbitrary subjective speculation and more or less objective knowledge.

The early positivists pushed these two advantages to unacceptable extremes by attempting to define them once and for all in a rigid way and then by using their absolute definitions as tools to bury all further metaphysical speculation. My own viewpoint is rather to use these advantages patiently and surely, gradually to clarify and to separate out the ultimately worthwhile speculations from those that are confused or otherwise unproductive. In this way, our philosophy, and in particular our epistemology, is done in the same pragmatic spirit as is science itself. Early positivism, by contrast, was a marriage of pre-scientific dogmatism with certain particular features of the then-current scientific method.

What I would like to do in this essay is to explore the relationship between certain ontological questions and certain epistemological ones. More precisely, I will be concerned with the question of how some traditional ontologies, notably Platonism, relate to epistemology done in the modern pragmatic spirit.

Platonism Revisited

It is well to recall at the outset that Plato's theory of ideas is a complete theory of knowledge involving at least two basic components, namely a metaphysics (i.e. a doctrine of what is ultimately real) and an epistemology (i.e. a doctrine of how to attain knowledge of what is ultimately real). Though these two components were considered by Plato to be inextricably bound together, they are nevertheless totally logically independent. Let us see briefly how this is so.

Plato held that the ultimately real objects were *forms* — eternal, ideal objects existing outside of space and time. The problem of knowledge, then, was reduced essentially to the problem of obtaining a clear and undefiled perception of these ideal objects (or, more precisely, as clear a perception of them as was humanly possible). Thus, whereas the individual might have to spend years learning how to interact with and to manipulate pragmatically the phenomena of the material world, such activity was at best a sort of 'trial practice', a discipline which helped to purify and clarify one's mental processes in preparation for the truly real knowledge that was to come only from a direct perception of the forms themselves. At worst, such pragmatic activity could even be a substantial hindrance to the achievement of the ultimate goal (witness the allegory of the cave).

Thus, whereas Plato certainly admitted that reasoning and sense experience were useful and necessary starting points on the road to knowledge, the ultimately most important mental faculty was intuition, for intuition, when properly purified and developed, was to give us direct perception of the forms. In other words, from this point of view reason and experience are lesser modes of knowing, gradually to be replaced by intuition as the seeker's inner eye is progressively opened by the discipline of knowledge he undergoes.

If I have correctly understood him in this regard, Plato believed that it is possible for the knower to arrive at a point where his intuition becomes the sole means of knowledge. This would be the stage at which the knower has acceded to the direct perception of the forms. Reason and experience applied to the material world are then as a ladder one has used to climb to the heights but which can now be thrown away.[2] In any case it is clear that, in Plato's conception, intuition becomes the dominant mode of knowledge, largely displacing the necessity for recourse to reason or experience related to the material (observable) world.

The logical independence of ontology and epistemology in Plato's total scheme can now be realized from the observation that the forms might well exist, and even be the cause of our capacity to interact successfully with the material world, without our being able ever to perceive them directly. In such a case, progress in knowledge would in no way remove the necessity for continual recourse to reason and sense experience. In short, the whole pragmatic method of modern science is wholly consistent with Plato's ontology. Plato's theory splits into the two logically independent halves of metaphysics and epistemology.[3]

Of course, it is easy to understand why Plato saw his theory of forms as an undivided whole. If one is convinced of the reality of the forms and of the fact that knowledge of them is the key to all knowledge, it makes sense to devise an epistemology whose ultimate goal is pure knowledge of the forms. For Plato, then, epistemology was a derivative of metaphysics. This is in direct contrast to the modern situation where we regard the epistemological method as given pragmatically by (successful) scientific practice but where the basic ontological questions remain unresolved. Plato resolved epistemological questions on metaphysical grounds. We are still trying to resolve ontological questions on epistemological grounds.

A note of caution should be struck here. It would clearly be wrong to regard scientific method and practice as themselves constituting some absolutely given Platonic ideal. As has already been stressed above, both scientific practice and our philosophical understanding of it are constantly evolving. But these considerations do lead us naturally to wonder whether the present state of our understanding of the nature of scientific method throws any light, one way or the other, on these ontological issues. Without falling into the anti-metaphysical dogmatism of the early positivists, but yet maintaining carefully our pragmatic epistemological stance, can we discern any direction either away from or towards something like Plato's metaphysics?

I think we can and I think that, on balance, the total direction is towards a Platonic ontology in some form or other. The remainder of this essay attempts to explain how and why I feel this way.

Platonism and Modern Scientific Practice

Given the nature of scientific method as currently understood and practised, what could be taken as evidence in favor of a Platonic ontology? The strongest positive evidence would, of course, be the direct and absolute perception of the forms themselves. However, we all know that such an experience of direct perception of the forms has not yet been forthcoming, nor has it been reported even by those scientists (such as the mathematician Kurt Gödel) who have believed most strongly in their existence. Scientific method is essentially the systematic and organized use of all of our mental faculties — intuition, reason, and experience — and successful science has always involved continual recourse to all of these faculties.[4] Nothing in the practice of science suggests that our intuition ever becomes sufficiently 'purified' to be independent of reason and experience.

In short, the accumulated experience of modern science, such as it now is, does not seem to suggest that direct perception of the forms, if they exist, is possible. Let us accept, then, as an established fact of our pragmatic epistemology (or as a working hypothesis, if you prefer), that direct and undefiled perception of whatever forms exist is not available to the human mind.[5]

In lieu of this greatest of all possible evidences for Platonism, what other kinds of evidence can there be? It seems to me that we can approach the problem in much the same spirit as we approach the question of the existence of material entities and forces which are not directly observable. We infer their existence by reasoning about observable configurations whose observed behavior seems unexplainable without them.

So we infer the existence of the force of gravitation because randomly dropped objects do not behave in a random manner; they all go perversely in a downward direction. This persistent deviation from presumed equiprobability leads us to construct pragmatically a non-probabilistic model of the motions of physical bodies in the presence of a large mass. Because this model turns out to be much more pragmatically acceptable and successful than its known alternatives, we feel obliged to acknowledge that there is something in the configuration of the phenomenon itself which allows this to be so. This 'something' is called the force of gravity.

Of course, what we call it is absolutely arbitrary. In fact, many of the properties we ascribe to it will also be arbitrary (or conventional) in various degrees, depending on the total model. Moreover, it is we, the knowers, who have conceived the model in the first place. Yet it would be wrong to say that the model itself is purely arbitrary or that it reflects nothing of the reality of the phenomenon external to our subjectivity, since so many other conceivable (and sometimes even plausible) models do not satisfy our pragmatic criteria to anywhere near the same degree.

What I am suggesting, then, is that the question of the existence of the non-material forms can be treated, pragmatically, in the same way as we treat the question of the existence of any theoretical entity, material or otherwise. I feel that it is possible to point to certain aspects of scientific method and practice which strongly suggest that there are 'somethings' out there beyond the purely material world of space and time, and that it is reasonable, pragmatically speaking, to identify these somethings with Plato's forms.

What I mean here by 'strongly suggest' is that the hypothesis of a realist ontology seems, in the case of these aspects of scientific method and practice, more acceptable on pragmatic grounds than the known materialistic alternatives. In particular, with regard to those aspects I will discuss, I consider that the materialistic alternatives either do not really explain satisfactorily what happens in practice, or else that the materialistic explanations are patently reductionist. I am obviously walking a tightrope between the law of parsimony on the one hand and the reductionist fallacy on the other.

Here, then, are the features of scientific method which I feel suggest a realist ontology:

1. The process of hypothesis and theory formation.
2. The discontinuity between fruitful and useless intuitions and notions.
3. The social nature of science, in particular the communicability of extremely abstract ideas and concepts.
4. The universality and applicability of many seemingly subjective ideas, especially with regard to the applicability of mathematics; the *a priori* nature of much of mathematics.

Let us discuss, in turn, each of these four aspects of scientific practice.

1. The paradigm of scientific method is that we start with experience on some level, and that we formulate a certain number of descriptive or observational statements which we call 'facts'. The process of amassing facts is generally a gradual, smooth process of carefully accumulating and tabulating observations, some of which will result from experimentation, i.e. from experiences we have deliberately provoked. Inevitably there comes a point in this process when we seek an 'explana-

tion' for the body of accumulated facts. We need a theory or hypothesis capable of relating the separate facts and welding them into a coherent whole.

Here the mental processes are reversed. Until now we have been interested in exploring how things, in fact, are. We now need to use our creative imagination in order to conceive how things might, in fact, be. But, as is now known, there are no rules for finding a fruitful hypothesis. The step of inductive reasoning is a discontinuous leap, a moving from a lower to a higher level of creative imagination and even of consciousness. This is true because there are generally a potentially infinite number of theories consistent with any given (necessarily finite) set of facts. In short, theory is under-determined by fact. Practically speaking, this means that, for any given phenomenon and at any given time, there is always more than one plausible or coherent explanation for our total experience of the phenomenon at that time.

Of course, once a theory is conceived, once a theoretically possible state of affairs consistent with the known facts is imagined, we do have a partial test of validity. We begin deducing as many consequences as possible from the theory, and in particular we try to deduce some new observational statements (singular judgements). These empirical consequences of the theory are 'predictions' rather than previously observed facts. If these predictions turn out to be true, then we have found a relatively fruitful explanation of the original set of facts. If it is the case either that some predictions turn out to be false, or else that hardly any testable observation statements are forthcoming (i.e. if most of the consequences of the theory are themselves theoretical), then we have, respectively, a false or a sterile theory.

The magnitude of the jump from the first level of amassing facts to the second level of hypothesis formation can be appreciated from the simple reflection that, unless some human mind succeeds in conceiving a fruitful hypothesis, the process of theory development will remain forever blocked, whereas we can go on amassing facts indefinitely. Since there are no rules for finding such hypotheses, the process of discovering them is highly dependent on human creativity. Moreover, considering the accumulated experience of scientific practice, we can observe that fruitful hypotheses are often based on precious few hard facts. From Newton's inverse square law to general relativity and quantum mechanics, scientific practice has presented us with the spectacle of increasingly complex and sophisticated theories inferred from a relatively low number of facts. The Michelson–Morley experiment, electron diffraction, the photo-electric effect, these are the empirical observations that have given rise to such fruitful and pragmatically useful theories.

Concomitantly, other areas of science such as biochemistry are still today virtually nothing more than an incredibly large mass of facts with no theoretical underpinnings anywhere near the degree of sophistication we find in, say, physics.[6] And even in physics there is an increasingly strong feeling of the need for some radical new insight, some new theory capable of providing a much more unified conception of the physical world.

All this is to say that we must take seriously the gap between the factual and the theoretical in science.

How does this bear on my thesis concerning a Platonic ontology? The ability of the human mind repeatedly to pick out a fruitful hypothesis with such accuracy, and often based on so few facts, strongly suggests that the mind has perceived an underlying form of which the material (factual) observable world is but a reflection. What is it that enables the mind to 'zero in' on relevant features of a phenomenon, discarding so many other apparently important aspects?

Let us recall in this connection that our theories and hypothetical constructs are *abstract* and *idealized* in the precise sense that they consciously (and unconsciously) neglect a myriad features of the phenomenon as it is perceived empirically. What is it, then, that enables the mind so often (though certainly not inevitably) to sift the relevant from the irrelevant and to avoid having to try laboriously the endless number of fruitless hypotheses before hitting a productive one? Again and again things seem to happen as if the mind has perceived some basic structure or form in a degree of clarity sufficient to enable it to develop a theory without much further recourse to observation of, or reasoning about, the material world. One reasons rather with one or more abstract structures or *models*, and it is the features of these models that guide further theorizing. If these models were essentially projections of human subjectivity, then it is difficult to see how they could guide us as surely as they do. The role these models play in scientific method is strongly suggestive of the objective existence of non-material forms underlying empirical phenomena.

2. Closely related to the discontinuity between fact-gathering and theory-creation is a second gap, namely between fruitful theories and false or useless theories. If the world is, in reality, very unstructured, one would expect an almost seamless, smooth continuum between correct and incorrect speculations. In other words, one would expect that if the differences between two formulations of an hypothesis were very small, then the difference in the resulting theories (i.e. the consequences of the respective formulations) would also be small. Sometimes this is indeed the case. But it is much more often the case that even minor perturbations in a viable theory lead to disastrous results rather than only to minor perturbations in the results.[7]

This recalls the overworn observation about the similarity between genius and idiocy. How far it is from our everyday, common-sense perception of the world to the sophisticated view of matter as little energy packets in relative equilibrium states whirling at tremendous speeds, with nothing in between! And those who have worked in a scientific field have all encountered instances of absurd but difficult-to-refute theories propounded by bright amateurs who nourish feelings of persecution by pointing to the analogy between their situation and the initial rejection of Galileo or Einstein by the establishment critics of the day.

Again, as with the example of theory-formation, things take place as if the material world were the imperfect reflection of precise structures and relationships which the mind succeeds in apprehending and expressing to a degree of clarity sufficient for valid prediction and control. If this is indeed so, then one can easily understand why even slight changes in a theory result in total falsity rather than in a slightly less useful theory.[8]

3. The third aspect of scientific practice which seems to suggest a Platonic ontology is the social nature of science, and in particular the communicability of extremely abstract ideas. Both from practice, as well as from the current state of learning theory and pattern recognition, we know how difficult it is to ascertain whether two different minds have the same concept of a given phenomenon. On the lower levels of communication involving simple abstractions related to demonstrable physical objects (e.g. color, size, shape), we can fairly easily accept a materialistic learning model based on association, conditioning, and the like. But no one seriously feels that we can explain highly abstract thought in this way.

In spite of the elusive non-materiality of abstract thought, sufficient communication does take place to enable science to continue its progress. Moreover, we know from scientific practice that progress in science depends on the creation of a community of understanding and a commonly shared framework of interpretation. No individual scientist, however brilliant, operates in a vacuum, in complete isolation from other scientists. Thus, the realist hypothesis that communication of abstract ideas is based on the common (though not necessarily identical) perception of a single abstract entity or form seems to be reinforced by modern scientific practice.

Also, the historical permanence and communicability of abstract ideas would seem to argue for some sort of Platonism. Some scientists have remarked, for example, that abstract mathematical ideas seem much less culture- and time-dependent than do literary or social ideas.[9]

4. The last feature of modern scientific practice that I would like to mention as suggestive of a Platonic ontology is the role of mathematics in the scientific enterprise. No one doubts that much of mathematics is prior to empirical experience. Yet over and over again mathematical ideas turn out to have wide applicability, giving immense predictive power. Modern science is shot through with what the physicist Eugene Wigner calls the 'unreasonable' applicability and power of mathematics. Why should mathematical theories formulated according to purely intrinsic criteria and abstract principles so often turn out to be powerfully applicable to the material world?

Again, we have a situation which suggests that the material world is an imperfect but highly approximate expression of some kind of pure form.

Since much has already been written about this by others, I will not belabor the point here. However, I would like to mention one variation of this theme due to Carl von Weizsäcker.[10] He points to the role of mathematics in providing simplifying, unitary, and basic formulations of extremely complicated phenomena. For him, physicists exhibit an almost mystic faith in the ultimate simplicity of the fundamental structure of the material world. Such a faith is justified not by our experience of the material world — a world which exhibits such bewildering diversity — but rather by our experience of *a priori* mathematical forms. Since the physicists' faith in simplicity has been substantially confirmed by the success of, say, quantum mechanics, one can only conclude that this basic intuition of ultimate unity and simplicity proceeds neither from our own subjective need for simplicity, nor from our experience of the material world (which does not justify it), but rather from our intuition of form itself. [11]

Materialistic Alternatives to Platonism

I am sure that my arguments in the foregoing will not have convinced any non-believers in the validity of Platonism. For even if one concedes that current materialistic models of scientific method are inadequate, one can still have faith that a successful materialistic model will eventually be found. Therefore, I do not intend to review a succession of non-Platonic alternative models of scientific practice. However, I would like to discuss one intriguing such model, presented by Professor Hans Mohr in his book *Structure and Significance of Science*. [12]

According to Mohr's model, the *a priori* in science is accounted for by the process of biological evolution. During the long period of time when the human nervous system was evolving, the survival pressures on individuals and populations were very great. There must have been, reasons Mohr, strong evolutionary pressures in favor of accurate thinking. That is, those genotypes whose central nervous system embodied a superior capacity to reflect accurately the structures of the material world would have been naturally selected ahead of those genotypes whose nervous systems were less well adapted in that regard. In this way, the essential structures of the material world gradually gave rise to corresponding structures in the physical human brain. The Kantian *a priori* is thus the result of experience of the material world which has been gradually accumulated by the human race as a whole, genetically encoded, and transmitted to each new generation. All knowledge begins with experience, in this view, but not necessarily experience on the part of the individual knower.

There is certainly some truth in Mohr's paradigm. There can be no doubt that many of our responses are pre-structured, especially the emotional ones which are closely linked to survival strategies (e.g. the mother–child bond). No doubt such pre-structuring exists also to some degree for logical and intellectual operations which, in spite of Piaget, must be regarded as still largely undetermined. However, I see at least two basic inadequacies in Mohr's model which are, for me, sufficient reasons to reject it as a reasonable explanation for our apprehension of structure in the framework of modern scientific practice.

In the first place, insofar as it explains anything, Mohr's model explains only how our brains were forced to develop a capacity to deal with the basic structure of the material world. It does not explain how or why there *is* structure. It does not explain how the immensely complex material world came to embody or reflect form or structure to a degree of regularity sufficient for our brains to apprehend it. His model does not really deal, then, with the problems dealt with in the Platonic model, namely the possible existence of non-material forms or structures which exist outside of and independent of the human brain.

There is a second inadequacy in Mohr's model which I find even more fundamental. Let us accept his hypothesis, which seems quite reasonable, that there were strong evolutionary pressures favoring accurate thinking. It seems clear that the type of thinking which would have had survival value during this primitive period of physical evolution would be thinking of an extremely practical and concrete sort, the kind that would prevent individuals from jumping off cliffs or eating poisonous substances. That this type of thinking would reflect some basic structures of the material world is quite clear. But the point is that the type of abstract, speculative thinking which is the basis of modern scientific theory-creation could hardly have had any positive survival value during this earlier period of evolution. One can hardly imagine selectivity in favor of quantum mechanics or general relativity.

In fact, the type of abstract thought involved in modern science is in many ways quite the antithesis of practical reason, requiring as it does the momentary suspension of virtually all practical considerations together with a certain emotional and physical disengagement from the immediate surroundings. Moreover, many of the modern scientific theories are quite counter-intuitive when viewed from the standpoint of practical reason and everyday experience.

During the more recent period of mankind's social evolution, only aristocrats, whose social situation allowed sufficient leisure and protection from the pressures of practical problems, were able to engage in such abstract mental activities. And these special conditions previously enjoyed by a privileged few have become general only in modern times and only in highly industrialized countries.

Indeed, the propensity for abstract, speculative thinking would have had a strongly negative survival value during the formative period of physical evolution. For an individual who had such a propensity and who attempted to indulge it would probably have been quickly eliminated in favor of his more practical-minded competitors. In contrast to this, one can imagine that a single flash of Platonic intuition would have sufficed to invent the wheel. (Does anyone seriously imagine that the wheel was invented by starting with an equilateral triangle and then gradually generalizing to regular n-gons as n approaches infinity?)

Thus, far from explaining the abstract and *a priori* in modern science, Mohr's paradigm seems ultimately to provide one more argument in favour of a non-materialistic model of scientific method and practice.

Conclusions

The above considerations sketch, in the broadest outlines, what I see as reasons for being open to the possibility of a realist ontology. Of course, there are problems with Platonism. If these forms or structures really exist, then what are they like? In mathematics, for instance, does 2 exist as an independent entity or do we construct 2 mentally (subjectively) from other existing forms – sets, for example? Among all the various foundational systems of mathematics, which, if any, more accurately describes existing forms? Are mutually reducible (i.e. logically equivalent) systems just different ways of articulating one basic form, or do there exist different universes of forms corresponding to different systems?

If any one system is a substantially more accurate description of existing forms than others, then this 'right' system should turn out ultimately to be more pragmatically useful in its various ramifications and extensions. But how far away is 'ultimately'? A true believer in a given system may be willing to wait much longer for the pragmatic justification than a nonbeliever. Yet believers have sometimes turned out to be right in spite of overwhelming opposition.

If we renounce any appeal to non-pragmatic epistemological criteria, as I have done in the present essay, then we can do nothing more than patiently develop and perfect our epistemology and abide by the answers it gives. However, the fact that pragmatic scientific practice is so strongly suggestive of Platonism must mean something important; the plausibility of Platonism should allow us to draw some kind of conclusion now concerning the pragmatic process itself.

I feel that, at the very least, it should spur us to be more open to the possibility of the existence of non-material entities as explanations for observed phenomena. We should not allow the materialistic ethos in which modern science has grown up to become uncritically transformed into an exclusive, dogmatic, materialistic philosophy elaborated in the name of science but in an unscientific spirit.

Contemporary exact philosophy is the offspring of early positivism. The early positivists were very sceptical of metaphysics, and they were right. Discussions about non-observable, non-material entities had too long been in the domain of arbitrary speculation which refused to submit itself to any objective or pragmatic criteria. Moreover, the early positivists were struggling to identify the essential features of scientific method and to build a coherent model of it. But we, having gained through experience a sense of sureness of our method, and having demonstrated by spectacular success its basic validity, need no longer fear the contemplation of that which is not directly observable.

The extreme scepticism of early positivism was healthy as an initial response to the emergence of modern science, but it is no longer necessary. We can, I feel, fearlessly apply scientific method in domains such as the religious and the spiritual, knowing that it will ultimately protect us from false imagination and, at the same time, open new and perhaps undreamt-of dimensions of human thought and experience.

2

Myths, Models, and Mysticism

Probably the single most significant characteristic of human nature is the individual's capacity for consciousness, or self-awareness.[1] This capacity endows every human being with a rich inner world of conscious internal states, a private world to which only he has direct access. The totality of a person's internal world constitutes his *subjectivity* and helps make of him a self-conscious, self-aware *subject* or *observer*.[2]

Let us agree to use the term *reality* to refer to the total sum of existence, i.e. to everything that exists, and the term *subjective reality* to refer to that part of reality made up of all internal human states – the sum of all human subjectivity. By *objective reality* we mean everything else besides subjective reality.[3]

According to these definitions, the chief feature of objective reality is that it exists outside the internal states of any human being. At least part of objective reality has a more concrete status: it is *observable* (*visible*) or *sensible* in that it can be directly

This essay is based on a paper contributed to the Symposium on Religion in the Modern World held at Ohio State University in 1983. This is its first publication.

perceived by all normally endowed human subjects by means
of their naturally given sensory apparatus and central nervous
system.[4] That part of objective reality which is not sensible
according to this definition will be called nonobservable or
invisible reality. Similarly, those internal, subjective states of a
given individual of which that individual is not himself aware
will be called *unconscious* states. The sum total of all
unconscious states is *unconscious reality*.

Let us summarize. By 'reality' we understand all that exists
– everything there is. By 'subjective reality' we understand
that part of reality which is wholly internal to one or several
human beings. By 'objective reality' we understand that part
of reality which is nonsubjective. Finally, that part of objective
reality inaccessible to human sense experience constitutes
'invisible reality', while that part of human subjectivity

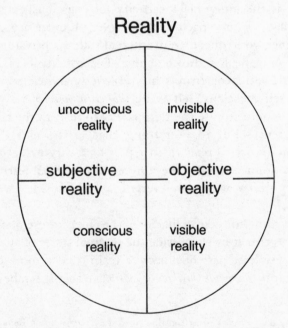

Diagram 1: The Basic Categories of Existence

inaccessible to any human consciousness is 'unconscious reality'.[5]

The self-explanatory diagram opposite illustrates these categories and their relationship to each other.

By a *phenomenon* we understand some portion (or part) of reality. Thus, a phenomenon can be wholly within any of the four separate categories, or it can involve various portions of some (possibly all) of them. A phenomenon can be thought of as more or less objective according to the relative portion that lies within objective reality. However, assessing degrees of objectivity is difficult since a phenomenon may contain huge portions of invisible reality (and thus be more objective than it appears) or huge portions of unconscious reality (and thus be more subjective than it appears).[6]

Theories

The main problem we face as self-aware subjects is how to obtain valid knowledge of the phenomena of reality. Though obtaining such knowledge will necessarily involve a certain amount of *experience*, i.e. interaction between ourselves and reality, the knowledge itself is internal to our subjectivity: any increase in knowledge will be reflected by some change in one's internal states, and usually in one's conscious internal states.

This basic epistemological situation is complicated by at least two things. First is the fact that, as knowing subjects, we cannot be wholly neutral in our search for knowledge. We have a number of needs that cry for satisfaction. These are partly tangible or physical — the need for food, shelter, and the like — and partly intangible or metaphysical — the need for meaning, a sense of purpose, a sense of self-worth. We do not, therefore, face reality either as self-sufficient gods or as infinitely flexible and adaptable creatures. If reality and the givens of our existence require things from us, we also have requirements and claims against reality, requirements that, in various degrees, we will not or cannot relinquish. We are therefore led to seek not just any knowledge but knowledge that is *useful*, i.e. that will help us fulfill our needs.

The second complicating factor derives from the realization that we do not know the extent of invisible reality and are therefore faced with an essential element of *mystery*. Moreover, our experience leads us to believe that invisible reality has significant influence on the behavior of visible reality.[7] In other words, invisible reality cannot be safely or conveniently ignored. We must, therefore, be forever alert to the possibility that we have seriously underestimated the invisible dimension of some phenomenon and thereby opened ourselves to potentially unpleasant surprises. The extent of the unconscious reality within us is also a mystery and may easily lead us to misjudge the internal resources available to deal with some particular life situation.

The confrontation between our needs and the mystery of invisible and unconscious reality creates tension and anxiety within us. We do not know whether we will be able to satisfy all our perceived needs, and we do not know whether we will continue to do so even when we are successful in the short run. At the same time, the existence of invisible and unconscious reality also helps to generate hope within us: even if we have so far been unsuccessful in fulfilling some individual or social need, we can always hope that we will in the future discover some heretofore unknown resource or power that can engender success. This can lead us to grasp at straws, to project our wishes onto reality, but it can also lead us to persevere to the point of success in the face of seemingly overwhelming odds.

In sum, obtaining valid and useful knowledge means obtaining a reasonably accurate mental picture or map of reality and matching that understanding with our needs in a way that allows us to fulfill them.[8] We will be unsuccessful if either our picture of reality is not sufficiently accurate, or else if the picture, though accurate, does not really help us satisfy any of our needs.

Indeed, we might say that the essential characteristic of human intelligence (which is a part of human subjectivity) is its capacity to mirror or to model phenomena. Subjective reality has the capacity to create or develop abstract, internal mental models of phenomena, and it is these mental representations that are really 'known'. They become the object of our scrutiny, contemplation, and social discourse.

Of course, our experience of objective phenomena, and our interaction with them, are crucial in allowing us to develop one or another internal picture of external reality. But once formed, the internal picture becomes, for the time being, the important thing. In particular, it will significantly influence the nature of our future interactions and experiences. It does not in itself change objective reality, but it changes the way we perceive reality, and this perception is what will largely determine our behavior in the immediate future.

Understanding how the mind makes internal pictures of visible reality is relatively easy (though we should not, perhaps, underestimate just how marvelous a process it is). But how do we deal with invisible reality? How can we make an accurate representation of something we have never observed? As it turns out, human subjectivity has an internal resource that seems to be designed for just such a task: it is the imagination – and more particularly, the *creative* imagination. By this latter term I mean our capacity to conceive of a configuration we have never actually witnessed. We can conceive of such a configuration as representing a possibly existing state of affairs. Let us agree to use the term *theory* to refer to any such imagined configuration. An observed configuration will be called a *fact*.

A theory, being the product of our imagination, may be rather fanciful and illusory, or it may correspond to some existing state of affairs. That is, the configuration we have conceived as representing a possibility may or may not correspond to some portion of reality, visible, invisible, conscious, or unconscious. A theory that describes accurately some phenomenon will be called *true* and, in particular, *true of* the phenomenon in question. We also recognize degrees of truth for theories. Thus, a theory may be said to represent a more accurate description of a given phenomenon than another theory without either theory being entirely true of the phenomenon.

Any phenomenon that a given theory purports to describe will be called an *interpretation* of the theory and will also be said to *interpret* the theory. Thus, a theory is true of a given phenomenon if that phenomenon interprets the theory and if the theory represents an accurate description of the phenomenon, and it is truer of the phenomenon than another theory having the same phenomenon as an interpretation if it represents a more accurate description of the phenomenon than does the other theory. We will also use the term *interpretation* to refer to the mental act or process of interpreting a theory.[9]

Another important characteristic of theories is their degree of usefulness. We will say that a theory is *useful* if it describes a state of affairs which would satisfy some important human need or needs. A useful theory describes a need-satisfying configuration. A useful theory is one we would like to be true. It may or may not be true. Thus, truth and usefulness, as here defined, are logically independent of each other.

It is our capacity for creative imagination that allows us to develop theories, but our motivation for doing so can be quite varied. On the one hand, we may be moved primarily by an extremely pure desire to determine as accurately as possible how some portion of reality functions: we may primarily seek truth. On the other hand, we may be urged to our creative task by the motivation of need-satisfaction; we may primarily seek usefulness.

However, a given theory may have been conceived as nothing more than an idle intellectual exercise with no pressing motivation and with little concern for its possible truth or usefulness. In this case, the process of theory creation appears as a particularly arbitrary and gratuitous subjective process. But there is a degree of arbitrariness inherent in any creative task, especially one like theory formation which requires so much flexibility of imagination. We would not expect such an activity to be definable by some rigid or strict set of rules. [10]

The point is that, however arbitrary the manner of a theory's conception, it may nevertheless turn out to be substantially true and/or useful. Truth and usefulness are qualities of the theory itself, not of the process that has generated the theory. This observation already gives a modicum of objectivity to theorizing since it allows us to assess the truth and usefulness of a theory without having to consider the often obscure mental processes which may have initially generated it.

Myths and Mythmaking

Once a theory is conceived, it may be articulated to others. Once articulated, it ceases to be the private intellectual property of its originators. The theory then becomes the object of whatever social processes are current within the society into which the theory is introduced.

Different societies may well have vastly different ways of processing new theories. At one extreme, a society may have a negative predisposition towards new theories, rejecting most of them out of hand. At the other extreme, a society may be inclined to favor new theories, eagerly embracing them and passing successively from one new theory to another. There are obviously many possible social configurations between these two extremes. For example, a society may cling strongly to certain kinds of theories while tending to reject others. Or a theory may appeal strongly to one segment of a society while appearing absurd or otherwise unacceptable to other segments.

In any case, the process by which a society or some segment thereof comes to adopt a theory may have very little to do with any attempt to test the truth of the theory. The process of acceptance or rejection may depend much more on whether or not the theory is perceived to be useful. A theory so perceived will be called an *attractive* theory (relative to the given society). A strongly attractive theory may gain widespread acceptance without much attention ever having been paid to its possible falsity.

Indeed, a sufficiently attractive theory may gain acceptance even if there is strong *prima facie* evidence against its truth. The point is that the processes by which theories gain acceptance in a society are social processes and, as such, do not necessarily have any intrinsic justification from an epistemological point of view.

Let us agree to use the term *myth* for any theory that has been accepted by a society or some significant segment thereof primarily on the basis of the theory's attractiveness. A myth may well be true, but it is not accepted because it has been tested and found to be so. It is accepted because it is perceived as an answer to one or more deeply felt social or widespread individual needs within the given society.

Nor does the falsity of a theory make it a myth. A false theory might well have been accepted only because, at the time, the evidence for its truth was perceived to be strong. What makes a theory a myth is the nature of the social processes by which the theory has come to be generally accepted, not the truth or falsity of the theory itself.

Let us apply the term *mythmaking* to any social process in any society by which a theory may be erected into a myth. Mythmaking thus describes certain kinds of social processes, namely those allowing a society that has them to adopt theories primarily on the basis of the attractiveness of those theories.[11]

It is one thing to define mythmaking, it is another to identify societies in which it has taken place or is taking place. Some might even argue that mythmaking as I have defined it has never occurred, at least on a large scale and in a systematic fashion. However, if we examine the history of human thought by means of those artifacts and written records available to us it appears that, during the millennia preceding the emergence of modern science, the exercise of the subjective faculty of creative imagination took place within a social context which tended to place few restraints on its inherent arbitrariness. Moreover, the theories that resulted from such spontaneous and undisciplined use of human imagination were quite frequently erected into myths. Not only did mythmaking abound, it was, in fact, the main thought paradigm of early civilization. It was the rule, not the exception.

The spectacle of societies willingly embracing theories whose truth is highly suspect may puzzle the modern mind impressed with its own scientific sophistication, but I think there are a number of fairly simple factors that explain rather well why mythmaking remained the thought paradigm of civilization for so long. First, there is the difficulty involved in testing the truth of highly speculative theories (e.g. the atomic theory of matter during the Hellenic period). It is, after all, much easier for me to know what I need or want to be true than to know what *is* true. My needs, especially my conscious internal needs, are part of my immediately accessible reality, whereas determining truth through multisubjective confirmation and verification is accessible only by means of highly complex forms of social organization and information exchange. Until such sophisticated social forms had time to evolve, the criterion of need-satisfaction (and thus attractiveness) naturally tended to remain dominant, if not absolutely determinant, in the social processes of theory acceptance.

A second factor is that there are usually many different plausible theories consistent with any given (finite) set of facts (see Note 10). This is the case even when the collection of facts is rather large, and when the collection is small, there are still fewer constraints on the set of logically possible theories.

As history progresses, humanity's fund of collective experience increases. Thus, early humans were faced with the task of explaining quite an extensive variety of phenomena, but based on relatively limited experience, i.e. with a relatively limited fund of accumulated facts on which to draw. It is therefore natural that this situation led to a veritable riot of speculative imagination. Since there was nothing in the mythmaking process itself which tended to limit such speculation, early societies were undoubtedly called upon to process an immense quantity of rather arbitrary theories. It was virtually inevitable that a certain number of these theories be accepted simply on the basis of their attractiveness. A society that processed new theories according to criteria of plausibility (i.e. probable truth) rather than attractiveness would not even waste time with a consideration of obviously fanciful theories.

The imagination, like any mental faculty, can be used either in a systematic and disciplined manner, or else spontaneously, sporadically, and arbitrarily. Even though a spontaneous and undisciplined use of the imagination may occasionally produce a true theory, one would expect that in the long run an organized and disciplined application of the imagination would be more likely to produce a greater number of true theories. Moreover, if originators of theories know in advance that their theories will be judged primarily according to truth criteria and only secondarily according to attractiveness, they will probably be more careful in theory formulation. In other words, a certain amount of preprocessing will occur in the mind of the theoretician before the theory is even articulated.

The social context thus operates on at least two levels with regard to the process of theory production and acceptance. It influences the way individuals go about the process of theory creation within the confines of their own subjectivity, and it also influences the way a theory is treated when once articulated.

All these considerations strongly suggest that, as long as mythmaking was the thought paradigm of society, the chances of evolving a substantial number of true theories were rather limited. An examination of the theories current in early civilizations seems to confirm this hypothesis. Most such theories are now perceived as obviously fanciful or at least completely discredited by subsequent experience. This fact explains why myths have come to be synonymous in the popular mind with mental fictions. In other words: a myth is not necessarily false, but it is awfully liable to be so.

Power Seeking and Conflict

Whenever we accept a theory, it ceases to be mere intellectual hypothesis for us. It becomes part of our worldview, of how we expect reality to behave. If a theory to which we adhere is challenged, our instinctive reaction is an aggressive and defensive one. We tend to cherish our accepted theories, to become emotionally attached to them. The greater our attachment, the greater will probably be our reaction if and when the theory is attacked.

An established theory can be challenged fundamentally in one of two ways. Either there is strong new evidence for its falsity, or else a competing theory gains acceptance among a rival group within society. If our theory is a myth, if we have accepted it primarily because of its attractiveness, then we may be able to resist invalidating evidence for quite a long time. Also, if an accepted theory is especially fanciful or *sterile* (making few affirmations about observables), then virtually the only way it can be challenged is by a competing theory.[12]

Thus, once myths are established, they tend to be maintained through defence against competing myths proposed

by rival groups in the society (or by rival societies). To defend
a myth by trying to demonstrate its truth may be epistemologically sound but socially ineffective since the myth was
accepted primarily on the basis of its attractiveness in the first
place. Thus the defence of one's myths ultimately reduces to
the material defence of the mythmaking community to which
one belongs.

In this way, competing myths tend to engender conflict
between mythmaking communities. The processes by which
one myth displaces another will tend to become identical with
the processes by which one group controls and dominates
another (thereby forcing the subordinate group to accept the
myth of the dominant one). Once this identification takes
place, individuals will naturally be led to seek power within
their society in anticipation of or in response to a threat to the
mythmaking community to which they belong. In sum:
mythmaking leads to power-seeking behavior and to conflict
within and among societies. The corroborating evidence
history offers for this thesis is so pervasive that one need hardly
do more than allude to it.

The Scientific Revolution: Model Building

I believe that the modern development of science, beginning
with the emphasis on empirical method in the European
Renaissance, and accelerating into the nineteenth and twentieth
centuries, represents a fundamental and irreversible transition
in the social and intellectual life of mankind. Whatever have
been its faults and philosophical excesses — mechanism,
positivism, behaviorism, reductionism — it represents a major
paradigm shift. It is a basic change in the mode by which a
significant (even if minority) segment of society tends to
process new theories: theories are to be processed primarily
according to criteria of truth and only secondarily according to
criteria of attractiveness.

Let us use the term *model building* to refer to those social and individual mechanisms that process new theories primarily according to criteria of truth. We are engaged in model building whenever we seek, individually and/or collectively, and by whatever means we have devised to that end, to determine whether or not a theory or theories are true, independently of the attractiveness of such theories. A theory that has been judged plausibly true according to these criteria is a *successful theory* or *model*.

Using this terminology, the scientific revolution may be said to represent a transition from mythmaking to model building as the fundamental thought paradigm of civilization. Indeed, we may give a broad, operational definition of science as the enterprise of model building. This defines science as a social enterprise, but without ascribing any specific content to it. The *scientific community* is that segment of society which is consciously committed to science. [13]

A model may, of course, be false, but only because we have been unable to detect its falsity by any of the means we have developed for testing truth. Moreover, our commitment to model building means that we will reject or modify the model whenever we have determined it is false, or at least sufficiently false to be no longer helpful in giving even a vaguely accurate picture of reality.

Just as a myth may be true, a model may be attractive and in fact useful. However, it is not accepted because of its attractiveness but rather because of its probable truth. Of course, nothing prevents us from concentrating our attention only on attractive theories, refusing even to test unattractive ones. But, no matter how attractive a theory, if we are committed to science, we will reject it if it fails to meet the criteria we have established for testing truth.

We should also bear in mind that the sincerity of our commitment to model building is not in itself a guarantee of success. We may apply our truth criteria very assiduously and carefully yet never come up with any successful theories, because our truth criteria can only be applied *a posteriori*, once a theory has been conceived and articulated. But the initial conception of a theory depends on a creative act of the human imagination, and we have no *a priori* guarantee that we will be sufficiently inventive to create a successful theory.

Nevertheless, if we look at the intellectual history of the last four hundred years, we can observe that the scientific community has succeeded in developing a number of highly validated models. Moreover, virtually all (if not indeed all) the successful theories we currently possess have been developed during this short four-hundred-year period. Thus, just as history seems to confirm that mythmaking does not produce many true theories, it also seems to confirm that model building does. The fact that the systematic attempt to process theories according to criteria of truth rather than attractiveness, even though practiced only by a minority segment of the total society, has produced so many successful theories in so short a time is a powerful confirmation of the (relative) efficacy of model building over mythmaking.

Truth Criteria

Determining the probable truth of a theory is not easy. Since a theory usually purports to describe a portion of invisible reality, we cannot check it directly by observation; at least we cannot check that part of the theory which deals with unobservables. Even checking the factual part can be difficult because of practical limitations on our capacity to observe remote, inaccessible, or small objects, or to make an exceedingly large number of observations. In the task of model building, we are thus faced initially with certain inherent, basic limitations, both of our minds and of our sensory apparatus. The realization of these limitations leads us rapidly to the conclusion that we cannot hope for absolute criteria for determining the truth of theories. We must accept that whatever tests and criteria we develop will be relative.

Nevertheless, the scientific community has evolved several criteria for dealing with theories, and a brief mention of some of them seems useful here. I will discuss three basic criteria — validity, adequacy, and simplicity.

We may assume that the process of building a theory begins with some form of sense experience which generates a certain number of facts or observations. In other words, theory construction begins by our attention being focused on some *concrete configuration*, i.e. some part of the visible portion of a given phenomenon. By a process of *idealization* we then make a mental map or picture of what we conceive to be the total phenomenon, visible and invisible. We *articulate* or *elaborate* the theory in the form of a body of propositions (or affirmations) that use both *abstract* terms (those referring to unobservable forces and entities) and *concrete* terms (those referring to observable forces and entities).[14]

The theory thus elaborated will usually make affirmations not only about invisible reality but also about the initially observed concrete configuration, and about other portions of visible reality as well. These latter affirmations are called *predictions* of the theory. Further statements are generated by the process of logical deduction, and new predictions result from the interpretation of these statements.[15]

If the concrete phenomena about which the theory makes predictions are accessible to us in our local space-time frame, then we can check by observation to see whether these predictions are confirmed. We can also interact with and manipulate our environment to try to provoke certain predicted configurations. These manipulative interactions with the environment are called *experiments*.

The whole process of checking the predictions of a theory

against portions of visible reality is the process of *theory validation*. If all the predictions of a theory that we have been able to check are confirmed, then the theory is *valid*.

The test of validity is a relative one for at least two reasons. Although we may have checked all predictions against known facts, there may yet come to light new facts that will contradict predictions of the theory. If this happens, then the theory will have to be either modified or abandoned. Also, though the process of generating predictions from a theory is a process of abstract logical deduction, it nevertheless evolves gradually. It is not done all at once. Thus, we may deduce tomorrow an affirmation that contradicts facts already known today. In this way, a theory may be invalidated by a process of deduction alone, without recourse to further observation.

With respect to the validity of theories, we are thus in a curious and somewhat uncomfortable position. It is possible to know certainly that a theory is false: if some of its predictions flagrantly contradict known and well-established facts, or if we derive a logical contradiction within it, then the theory cannot be true. But even if we have been able to check all known predictions against all known facts, and have derived no logical contradiction, we cannot exclude absolutely that new facts and/or new predictions may, in the future, spell the doom of the theory.

Nevertheless, the longer a theory persists and satisfies the criterion of validity, the more confidence we can reasonably have in its probable truth. This is especially the case if the theory is *rich*, i.e. if it generates many testable predictions. A theory, most of whose consequences are themselves theoretical, is a sterile theory. Thus, a sterile theory will make many affirmations about invisible reality and few or none about visible reality. Many myths are sterile theories.

As previously mentioned above, there are usually several different logically possible configurations consistent with any given set of facts. We are therefore often faced with the task of deciding which of several valid theories is plausibly better. Adequacy and simplicity are comparative criteria that help us make such decisions.

Adequacy refers to the amount of visible reality explained by a theory. Even though a theory starts out to explain only some very particular phenomenon, it may end up explaining quite a bit more. For example, Newton's theory of gravitation sought to explain how and why unsupported objects fall to earth. It ended up explaining not only that but the motions of the planets and a lot of other things as well.

Simplicity (traditionally called 'Occam's razor' or the 'law of parsimony') refers to the relative complexity of theories. It represents our desire to avoid gratuitous assumptions and arbitrary speculations. Given two valid and equally adequate

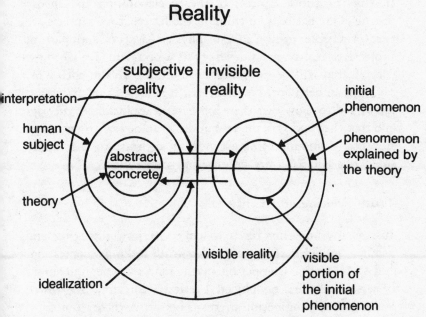

Diagram 2: The Process of Theory Construction

theories, we will choose the simpler of the two, i.e. the one that posits the existence of fewer (or less complicated) nonobservable forces and/or entities.

There is a certain trade-off between adequacy and simplicity. We will accept a substantially more complicated theory if it is also substantially more adequate. This inverse relationship between simplicity and adequacy can be seen, for example, in the transition from Newtonian to Einsteinian theories of gravitation.

The diagram overleaf summarizes the process of model building by illustrating the basic relationships between the essential components involved in the process of elaborating a theory. Here, we do not show the division of subjective reality into its conscious and unconscious domains.

Our discussion of truth criteria and the above diagram are certainly far from complete in their treatment and representation of the subtle process of theory elaboration and theory testing. One has only to think of the fact that we humans are part of visible reality in our physical aspects and part of subjective reality in our mental aspects to realize how complicated will be the process of building a valid theory of, say, human behavior. Nevertheless, it is hoped this discussion has served to show that there are criteria applicable to theories and that the process of model building can be carried on with a significant amount of objectivity, in contrast to the highly subjective and arbitrary nature of mythmaking.

Truth Seeking and Unity

Just as mythmaking tends to engender power seeking and conflict, model building engenders truth seeking, cooperation, and social unity. Competing myths tend to generate conflict because there are no objective criteria for deciding which among several competing myths is better. A theory is a myth because it is perceived as an answer to one or more needs of the mythmaking community, and the needs of different myth-making communities may be (or may be perceived to be) conflicting or even irreconcilable.

Models may also be mutually (logically) incompatible, and it may well happen that rival models of the same phenomenon come to be championed by different segments of the scientific community. Nevertheless, to the degree that these scientific subcommunities are faithful to their respective commitments to the model-building process, they will resist indulging in power-seeking or dominance-seeking behavior in their attempts to resolve the incompatibility between rival models. Committed to finding the simplest and most adequate among all valid, known models, these subcommunities will seek rather to collaborate in a common effort either to decide in favor of one model or the other, or else to find a new model superior to all. In fact, the greater the tension of incompatibility between two relatively successful theories, the greater will be the motivation on both sides to resolve the tension by establishing a more successful, integrated model of the phenomenon being studied.

Such, of course, is the ideal, and if things have not always worked out that way in practice, they have worked out that way often enough for science to have maintained itself as an ongoing, flourishing, and indeed growing enterprise. Moreover, a number of blatant attempts to subvert the integrity of the model-building process (e.g. Lysenko in biology or C.R. Burt in psychology) have been ultimately frustrated through detection and subsequent denunciation.

It is not just the logical incompatibility between rival theories that may lead to social conflict but also the degree of emotional attachment each scientific subcommunity may feel for a particular theory. This emotional attachment can create the need to defend the theory independently of its merits, thereby generating emotional pressure to abandon or compromise the model-building process and to revert to myth-

making. In this way, a model can be transformed into a myth
and ultimately defended by the same manipulative, power-
seeking techniques society has had so much practice developing
during the long millennia when mythmaking was the thought
paradigm of civilization.[16]

The scientific community functions within the community
at large, and scientists share the tangible and intangible needs
of humankind. It is not reasonable to expect that integrity of
commitment to model building and truth seeking can be
indefinitely maintained on a large scale in the face of powerful
social inducements and pressures to the contrary. Society can
thus undermine or destroy the model-building enterprise
either by threats and punishments, which create enough fear to
discourage commitment to truth seeking, or else by material
and social rewards that corrupt the scientific community.

All of this is to say that the scientific enterprise is socially
fragile. We stand constantly in danger of its being destroyed
or undermined by sufficiently negative social configurations
and processes.

Science and Religion

Science, as defined and described above, has no specific
content. It represents a method or way of pursuing the
enterprise of seeking knowledge, particularly knowledge about
invisible reality. Specific sciences, disciplines, and techniques
result from the application of the model-building approach to
specific phenomena. But one can approach the study of these
same phenomena from the point of view of mythmaking and
forget about scientific method altogether. Thus, it is not the
phenomenon studied, the knowledge sought, or the questions
asked that determine whether or not a given discipline is
scientifically legitimate, but the method or approach used.
Science is defined by its method.

Religion, however, is defined by its content and goals.

Following Karl Peters, we may give a heuristic definition of religion as the enterprise of seeking knowledge about what is ultimate in invisible reality, especially what is ultimate in relationship to human life and experience. [17] Religion therefore considers such issues as the nature and scope of human consciousness, the possibility of life after physical death, the meaning of pain and suffering, the existence and/or meaning of good and evil, or the possibility of transcendent experience, i.e. of communing with or experiencing subjectively that which is ultimate in invisible reality. [18]

Religion, so defined, is certainly amenable to the model-building process and is therefore scientifically legitimate. However, we know as an historical fact that the same four-hundred-year period which has witnessed the growth and development of science has also witnessed an intense conflict between established religion and an increasingly established science. Let us consider briefly some of the factors that may have contributed to the generation of this conflict.

Because the questions asked and the knowledge sought by religion are so fundamental, vital, and universal, religion seems to have been among the first human activities to be organized socially. Even today the institutions associated with organized religion are, for the most part, traditional ones growing out of a long history. Religion is thus an ancient endeavor, and one that flourished mightily during the long period when mythmaking was the dominant thought paradigm of human society. Moreover, religion was the most powerful, established social force when the European Renaissance scientific revolution occurred. Since, as we have already noted above, mythmaking engenders power seeking and conflict, it was natural (though not necessarily inevitable) that religion enter into conflict with the new science. Even though most Renaissance scientists were deeply religious men, and, in fact, often religiously motivated in their scientific under-takings, they were nevertheless the carriers of a new paradigm of model building and were consequently rejected by the religious establishment, which was so largely based upon and so thoroughly committed to the mythmaking paradigm.

After this initial breach between traditional religion and nascent science, the model-building process grew and developed independently of religion, while many exponents of religion maintained a dogmatic, obscurantist, and reactionary attitude towards science. As a result, the separation between religion and science tended to widen and to solidify. This has had unfortunate consequences for both religion and science.

One of the main consequences of the four-century-long conflict between established religion and established science is that, in the minds of many, religion has come to be identified with mythmaking itself. Secular science is perceived to be the expression, *par excellence*, of model building while religion is perceived to be the expression, *par excellence*, of mythmaking. When viewed in this way, religion is seen as inherently and intrinsically bound up in mythmaking and therefore not scientifically legitimate as a knowledge-seeking enterprise. This perception has led many of the brightest minds and most sensitive spirits to turn away from religion, thereby depriving religion of vital insights these minds might otherwise have contributed to it. Adherence to religious belief has now come to be viewed primarily as an emotional attachment to certain comforting illusions – a collective neurotic mechanism for dealing with the difficulties of life.

For science, the main result of its conflict with religion has been science's overnarrow concentration on the exhaustive study of certain material phenomena, coupled with an almost total indifference to the more global, universal questions asked by religion. The model-building process has never been seriously applied to religion and religious questions because the scientific community long ago abandoned the consideration of such questions, preferring to leave them in the hands of traditional religious dogmatists. Science has come to mean not just the model-building process but rather the particular application of that process within certain limited domains of inquiry and from a certain narrow viewpoint. Even to ask religious questions or to seek knowledge about ultimate reality is often perceived as *per se* unscientific. In this way, the practice of science has become wedded to a dogmatic philosophical materialism, a materialism that is tacitly and erroneously considered to be inherent in science itself.

As a result of the centuries-long adversarial relationship between science and religion, we now have highly successful models of certain limited material phenomena on the one hand (the fruit of four hundred years of scientific activity) and only vague and arbitrary metaphysical speculations about ultimate questions on the other.

Of course, we have no *a priori* guarantee that the application of model building to religion will result in successful religious theories, any more than we had any such *a priori* guarantee that model building would be successful in the first place. But it is certainly unfair (and unscientific) to assert dogmatically that model building must fail when applied to religion without having seriously and systematically attempted it. Moreover, if we have taken several hundred years to develop successful theories of certain limited and relatively accessible material phenomena, it is reasonable to suppose that we may take at least that long to succeed in the obviously more complex and difficult domain of religion.

In contemplating the application of model building to religion, one is tempted to start by discarding *in toto* most of traditional religion, since the theologies of these religions are so permeated with mythmaking. It seems very difficult to distill the true from the false, the valid from the invalid, the real from the imaginary in traditional religion.

However, starting from scratch in applying model building to religion is certainly a daunting prospect. It is also tinged with the arrogant assumption that there is nothing of scientific value to be gleaned from thousands of years of collective human experience, including the thought of some of the deepest thinkers history has produced.

In this regard, there is an interesting theory of the history of religion which, if true, would enable us to apply model building to religion without such a radical and wholesale rejection of traditional religion. This theory, called *progressive revelation*, is discussed in detail in another essay in this book[19] and it will suffice here to recall its chief features.[20]

If we examine the history of religion, we can see that there are at least two different social processes which have often been included under the name of religion. One process is the generation of taboo systems by society. This process is gradual and anonymous, and frequently leads to a 'common-denominator religion' that enshrines mainly those elements acceptable to a majority of the society. I would include some (though not all) forms of animism, fetishism, and shamanism in this category.[21]

The second category of religious processes is represented by those religious systems founded by a single charismatic teacher and leader, a prophetic figure who appears, suddenly and discontinuously, as a revolutionary within his society. The teachings this figure gives are often at variance with accepted tradition, and the teacher himself is usually persecuted (and frequently killed) because of the reaction his ideas provoke in certain segments of the society. Nevertheless, he manages to attract a modest group of followers, and from this initial group develops a community of believers who accept the new teachings and strive to implement them. The founder of such a religion presents himself as a 'prophet' or God-inspired revealer of true theories about invisible reality. He makes certain promises (a covenant) and predictions (prophecies), and invites his followers to use them to test the truth of his theories. He may also write or dictate a book, which helps protect his teachings from the ravages of oral tradition.

Religions of this second type are variously called 'prophetic religions', 'higher religions', or 'revealed religions', and the above sketches some of the facts on which the theory of progressive revelation is based. The central hypothesis of this theory is that the founders of the higher religions were indeed (as they claimed to be) revealers of true theories about invisible reality. Thus, whereas common-denominator religions may well be expressions of pure mythmaking, the revealed religions are not. They represent a different kind of social process, one that generates social change and challenges existing institutions. It is a process initiated by the creative genius of one insightful and inspired individual. According to the theory of progressive revelation, Moses, Jesus, Buddha, Muḥammad, and Zoroaster are among the revealers of true religious theories.

Of course, even if we grant the central hypothesis of the theory of progressive revelation, and acknowledge that the founders of the higher religions were revealers of true theories about invisible reality, the fact remains that mythmaking was the dominant thought paradigm of the societies which received the teachings of these spiritual geniuses. Their teachings were therefore processed according to the paradigm of mythmaking rather than that of model building.

Instead of putting the new theories to the test of truth as suggested by the founder, societies proceeded to deify or to idolize the founder (putting him comfortably out of reach), to embellish, elaborate, and distort his teachings in ways that satisfied the particular needs and desires of the society in question,[22] and to establish authoritarian, priest-dominated religious communities which eventually became the new establishment. Powerful metaphors and analogies used by the founder to explain the structure of invisible reality were transformed, through literal interpretation, into dogmatic theological pronouncements that individuals challenged on pain of ostracism or even death.

Still, within each revelatory tradition there has been a

(minority) fragment of believers who have responded directly to the message of their founder, understood at least partly the deeper implications of that message, and validated through personal experience and social action its truth. Thus, if the theory of progressive revelation is correct, we do not have to start the model-building process in religion from absolute nothingness. We have at least those original teachings (e.g. the Qur'án of Muḥammad) which have been preserved and whose historical authenticity has been reasonably validated. We also have the collective thought and experience of those who have responded seriously to the original message and who have left a record of their earnest attempts to implement their understanding of it.

The theory of progressive revelation, in the form sketched above, was first articulated in the late nineteenth and early twentieth centuries by Bahá'u'lláh (1817–1892) and his son and designated interpreter 'Abdu'l-Bahá (1844–1921). Bahá'u'lláh, who founded the Bahá'í Faith, taught that the progressive nature of the phenomenon of divine revelation was due not only to the relativity of truth and the necessity for the elaboration of ever more adequate true theories of invisible reality but also to the need to provide an adequate basis for social organization. He explained that the direction of social evolution was towards the organization of society on progressively higher levels of unity, culminating in the (as yet to be achieved) unity of the planet itself in one coherent social system.

According to Bahá'u'lláh and 'Abdu'l-Bahá, productive and satisfying human relationships, and a just and efficient social order, are too complex and too fragile to be founded on inadequate or invalid theories of human nature (or of reality in general). The social purpose of religion, they affirm, is to provide true theories of invisible and unconscious reality which are adequate for building a stable and progressive society, and which, at the same time, furnish individuals with a true and accurate understanding of their own internal reality.

In this regard, it is most interesting to note that the Bahá'í Faith is one of the few (perhaps the only) major religious systems which came into existence after the scientific revolution of the Renaissance had taken place. Moreover, both Bahá'u'lláh and 'Abdu'l-Bahá repeatedly stressed that the model-building process is the only one capable of leading man to discover and/or correctly process true theories. For example, speaking in Paris in 1911, 'Abdu'l-Bahá stated:

Consider what it is that singles man out from among created beings, and makes of him a creature apart. Is it not his reasoning power, his intelligence? Shall he not make use of these in his study of religion? I say unto you: weigh carefully in the balance of reason and science everything that is presented to you as religion. If it passes this test, then accept it, for it is truth! If, however, it does not conform, then reject it, for it is ignorance![23]

Concerning the conflict between religion and science, he affirms:

All religions of the present day have fallen into superstitious practices, out of harmony alike with the true principles of the teaching they represent and with the scientific discoveries of the time . . . The outcome of all this dissension is the belief of many cultured men that religion and science are contradictory terms, that religion needs no powers of reflection, and should in no wise be regulated by science, but must of necessity be opposed, the one to the other. The unfortunate effect of this is that science has drifted apart from religion, and religion has become a mere blind and more or less apathetic following of the precepts of certain religious teachers, who insist on their favourite dogmas being accepted even when they are contrary to science.[24]

Regarding mythmaking and model building, 'Abdu'l-Bahá has, in another context, stated: 'If religious beliefs and opinions are found contrary to the standards of science they are mere superstitions and imaginations . . .'[25] This idea is more completely articulated in the following passage, again from 'Abdu'l-Bahá:

Reflect that man's power of thought consists of two kinds. One kind is true, when it agrees with a determined [reality]. Such conceptions find realization in the exterior world; such are accurate opinions, correct theories, scientific discoveries and inventions.

The other kind of conceptions is made up of vain thoughts and useless ideas which yield neither fruit nor result, and which have no reality. No, they surge like the waves of the sea of imaginations, and they pass away like idle dreams.

In the same way, there are two sorts of spiritual discoveries. One is the revelations of the Prophets, and the spiritual discoveries of the elect. The visions of the Prophets are not dreams; no, they are spiritual discoveries and have reality . . .

The other kind of spiritual discoveries is made up of pure imaginations, but these imaginations become embodied in such a way that many simple-hearted people believe that they have a reality. That which proves it clearly is that from this controlling of spirits no result or fruit has ever been produced. No, they are but narratives and stories.[26]

To summarize: from the Bahá'í viewpoint, there is no essential or basic opposition between religion and science. Religion is a knowledge-seeking enterprise and, as such, can do no better (nor worse!) than to apply the model-building process to its particular domain of inquiry. The Bahá'í theory of progressive revelation, if correct, provides the basis for a rational understanding of the relationship between modern scientific developments and traditional religious systems based on the teachings of the great religious founders of history.

A concomitant to the application of model building to religion is the necessity for the scientific community to renounce its irrational attachment to dogmatic philosophical materialism. This process also seems to be gaining ground, and a number of recent books by practicing scientists have stressed the inadequacy and reductionistic character of materialism in science.[27]

Mysticism

Mysticism may be defined as the attempt to obtain direct experience of invisible reality and, more particularly, of what is ultimate in invisible reality. Such an experience, if it exists, is necessarily an inner, private subjective experience since, by definition, anything in invisible reality is not directly accessible to physical observation or sense experience. Thus, religion is a knowledge-seeking enterprise, while mysticism is an experience-seeking enterprise.[28]

Seeking knowledge of invisible reality leads to the basic problem of distinguishing between true and false theories. In a similar way, seeking experience of invisible reality leads to the problem of distinguishing between possibly valid experiences, on the one hand, and self-generated illusion, on the other. It is only recently, historically speaking, that mankind has evolved a reasonably efficient technique for obtaining knowledge of invisible reality: the scientific method (i.e. model building). No such comparable technique has yet been developed for obtaining experience of invisible reality. We are thus liable to have feelings of malaise towards traditional mysticism that are comparable to (and perhaps even stronger than) those we feel towards traditional religion. Just as we feel tempted to discard much of traditional religion as mythmaking, we are similarly tempted to discard most of traditional mysticism as illusory — the product of psychological suggestibility and over-active imagination.

Skeptical attitudes towards traditional mysticism are perhaps somewhat more warranted than are similar attitudes towards traditional religion. In the case of religion, there is a certain degree of objectivity deriving from the large number of participants in the religious enterprise. The historical processes affecting the birth and development of a given tradition can be studied and analyzed. Moreover, many of the propositions and affirmations contained in religious theories have empirical content and can therefore be tested experientially to some extent. While this is also true with some branches of mysticism, which make predictions about the empirical effects of certain techniques of spiritual discipline, the mystic tradition as a whole seems much less objective in spirit than religion.

Indeed, much of traditional mysticism and its modern

counterpart, existentialism, tends to glorify human subjectivity and its richness, in contrast to the perceived banality or sterility of everyday sense experience. From this point of view, the very essence of true mystic experience is that it cannot be articulated, communicated, or modeled.

Modern psychology, which has given us the powerful theoretical construct of the unconscious mind, has also documented just how real the products of human imagination can seem to those who generate them. If, as the cliché goes, there is a fine line between the genius and the idiot, then there is an even finer line between the mystic and the schizophrenic. Immersed in the labyrinth of his internal mental process, harboring a disdain for the flow of everyday sense experience (from which our sense of reality and identity is largely derived and on which our theories are largely based), the mystic cannot help but lose his way more often than not.

Yet the fact that the successful pursuit of mystic experience may be difficult does not, in itself, invalidate mysticism, any more than the enormous difficulties inherent in model building invalidate science. It only means that we have to be at least as careful, if not more so, in our attempts to experience invisible reality as we were in our attempts to know invisible reality. Indeed, there is no reason why some of the techniques of scientific method cannot be applied in mysticism as in religion, and I would like to mention a few ways in which I see the relevance of model building to mysticism.

In the first place, one of the principal goals of the modern science of psychology is to build adequate and valid models of internal human functioning. Science thereby enables us to improve our understanding of subjective reality, both conscious and unconscious. The more accurate our understanding of our internal reality, the more capable we become of distinguishing between what is self-generated and what is possibly other-generated in our internal experience.

Second, to the degree that we have highly validated models

of invisible reality, we have some basis of knowledge on which to judge what we may reasonably expect the experience of invisible reality to be like. In particular, it seems reasonable to expect that genuine mystical experience should be universally accessible and open to multi-subjective confirmation and validation. If a mystical experience is an experience *of* something other than one's own self, if it is a response to some outside invisible force or entity, then presumably there will be features common to everyone's experience of that force or entity, just as there are features common to everyone's sense experience of the same physical object. Such an expectation stands in significant contrast to the point of view of existentialists and some mystics who glorify the chaotic, unpredictable, irrational, intensely private, and incommunicable nature of mystic experience.[29]

Of course, none of this gives us absolute criteria for assaying mystic experience, but then we cannot reasonably expect that such absolute criteria will be forthcoming here when they do not even exist for ordinary sense experience.

Mysticism and Modern Science

Recent years have witnessed a rebirth of interest in mysticism, especially that of the Eastern tradition. Some authors have held that mysticism represents a method of knowledge parallel or complementary to that of science. Others have held that modern science, in particular theoretical physics, bears an increasing resemblance to certain forms of mysticism.[30] Applications of systems theory to building models of complex physical systems has led to a 'holistic epistemology' which, it is said, is quite similar to the mystic goal of experiencing reality as an undifferentiated whole. It is clear that authors who write in this vein anticipate some kind of philosophical *rapprochement* between modern scientific and traditional mystical (and thereby religious) categories of thought and experience.

While agreeing with some aspects of this thinking, I find myself out of sympathy with its fundamental thrust. I would therefore like to use the general framework developed in this essay to explain why I feel as I do.

To begin with, it is certainly proper and important to underscore the severe limitations of the materialistic-reductionistic view of the nature of science. There is an increasingly widespread rejection of materialism and positivism in many quarters, and that is undoubtedly healthy. Insofar as exposing the fallacies of this overnarrow philosophy of science is the goal, I am quite in accord. But what may be fairly regarded as unhealthy exaggeration is the rejection of the paradigm of model building itself. For example, when one speaks of overcoming the subject—object dichotomy or of a close analogy between quantum mechanics and ancient mysticism, it is legitimate to wonder whether such a move is not regressive rather than progressive.

In this connection, we should recall that traditional mysticism developed during the same period when myth-making was the thought paradigm of civilization. If people could easily convince themselves that certain unvalidated but attractive theories were true, they could just as easily convince themselves that certain intense inner experiences were nothing else but the Voice of God speaking to them. Of course, they may not always have been wrong any more than myths were always false. But they were liable to have been more often wrong than right.[31]

I sense, in the current fascination with mysticism, a certain nostalgia for the primitive days of undifferentiated wholeness, before science gave us many of the painful distinctions with which we must now learn to live. For historical reasons (mainly deriving from its conflict with religion) science has, to a considerable extent, overprivileged reason and analysis at the expense of intuition and synthesis, even though, as we have seen, no such imbalance is inherent in model building itself.

There is consequently a great hunger for wholeness, for a new global vision, abroad in the world today. However, the proper response to this legitimate need is not a regression to the undifferentiated wholeness of mythmaking and traditional mysticism but rather a progression to a new synthesis. Some reflections on the form this new synthesis may take constitute the next and concluding section of the present essay.

The Age of Synthesis

We have examined the transition from the prescientific to the scientific age as a transition from the paradigm of mythmaking to the paradigm of model building. From this point of view, one might have expected that the widespread adoption of the model-building paradigm would solve the basic problem of human existence, i.e. the problem of obtaining a knowledge of the phenomena of reality adequate to human needs. However, this has clearly not occurred. While we have gained tremendous power to manipulate our physical and social environment, we have, at the same time, engendered weapons of war of unimaginable destructive power and developed a highly materialistic and extremely fragile society that enshrines massive social and economic injustice as well as a deep and widespread sense of personal and spiritual alienation.

Science (or at least what we have done with science) has made us neither happy nor secure. As Carl Jung once expressed the idea: through science, and the use we have so far made of it, we have conquered nature; but we have not yet understood or conquered our own nature. The present moment in history may be fairly characterized by the fact that we now have the certain knowledge of how to destroy ourselves but only the vaguest, unsupported speculations about how to prevent such destruction.

If the analysis in the preceding sections of this essay is reasonably correct, the failure of science to solve the problems of human existence can be attributed in significant measure to the fact that the practice of science has, for historical reasons, become linked to a narrow materialistic-reductionistic philosophy, thereby undermining the application of model building to certain legitimate and vital areas of human concern. Nevertheless, this is not quite all the story. Both mythmaking and model building are social processes and, as such, cannot be taken as metaphysically (or even epistemologically) ultimate. Our identification of science with model building, and our consequent interpretation of the scientific revolution as basically social rather than metaphysical, leads us directly to the question of the nature of social processes themselves. Since it is a social and historical process that has produced science in the first place, we must try to understand the basic mechanisms and forces underlying such processes if we are to build an accurate model of how such (future) changes (may) occur.

In short, we stand in need of a valid and adequate theory of history, one that allows us to understand and explain the emergence of science from a broader perspective. The basic scheme we present is closely linked to the theory of progressive revelation, which was briefly discussed in the section 'Science and Religion' above. Let us call it the *organismic* theory of history.[32]

According to the organismic view, the human race constitutes an organic unit that is involved in a collective growth process analogous to the growth process of a single individual within a society. In the same way that an individual begins life as a helpless infant and achieves maturity in stages by gradually accumulating an increasingly sophisticated complex of abilities, so humanity's social life has moved from its primitive beginnings through a succession of stages that are leading towards an ultimate, stable configuration representing the maturity of the human race, the culmination of social evolution. Though the successive stages in humanity's collective evolution may be analyzed in various ways, they resolve into three basic periods, which we will call primary integration, differentiation, and secondary integration. These correspond roughly to childhood, adolescence, and adulthood in the life of an individual.

Primary integration represents the long period when mythmaking was the basic thought paradigm of civilization. It is characterized by a relative lack of a sharply differentiated human awareness. In this stage, people tended to perceive themselves as part of a whole, as one with nature and a preestablished natural order into which they fit. This is strongly analogous to the child's perception of himself as undifferentiated from his family and immediate environment.

In the same way, the unbridled use of creative imagination, which gives rise to rampant mythmaking, is analogous to the child's use of imagination in his attempt to relate to the world. For the individual, the undisciplined use of the imagination in childhood must precede its later disciplined use; otherwise, the imagination will never be sufficiently developed in adulthood to conceive of the relatively complex theories needed to build adequate models of reality. Thus, the organismic theory of history views the long period of mythmaking in humankind's collective life not as an unfortunate accident or a regrettable waste of time but rather as a healthy and necessary preparation for the later stages of collective development (including the stage of model building itself).

The organismic view of history is thus a frankly teleological one. According to this theory, the basic periods in the collective life of mankind are not accidents, nor are they purely the result of a social 'natural selection'. They are viewed as goal directed.

The second basic stage in the process of social evolution, that of differentiation, is analogous to adolescence in the life of the individual. For the individual, adolescence is characterized by mature physical development coupled with relatively immature emotional and spiritual development. The essential task confronting the adolescent individual is that of defining his or her identity through the achievement of spiritual and emotional maturity, a task the adolescent pursues by exploring all the ways he or she differs from others. Thus, through competition and conflict, analysis and criticism, the adolescent forges self-awareness.

According to the organismic view of history, the physical development of the collective human organism is represented by advances in science and technology, while emotional and spiritual development are represented by the nature and quality of human interactions and by the integrity of the social fabric generally (for example, by the relative degree of social and economic justice that prevails within society). Thus, the adolescent stage in the collective life of humanity is characterized by a relatively high degree of scientific and technological achievement, coupled with relatively immature forms of social organization and human interaction.

This characterization of collective adolescence describes rather well the condition of the modern world. In particular, the relatively high level of scientific and technological attainment, which is so pervasive a feature of modern society, has its origins in the seventeenth-century European scientific revolution, i.e. in the transition from mythmaking to model building. Thus, from the point of view of the organismic theory of history, this transition represents the passage from the collective childhood to the collective adolescence of humankind. It has resulted in the predominance of our analytical and critical powers and in the development of a keen, indeed painful, self-consciousness.

Like the theory of progressive revelation, the organismic theory holds that the general direction of human social evolution is towards greater complexification through the organization and reorganization of society on progressively higher levels of unity. Each higher level of unity implies a greater degree of specialization of social and economic roles, as well as a correspondingly greater degree of interdependence and mutual trust among the differentiated parts of society. Of course, this collective growth pattern in our history is not anything like an uninterrupted, linear ascent. Clearly there have been ups and downs, fits and starts, successes and failures. Nevertheless, we know that there was a time some six to ten thousand years ago when social organization was extremely crude and limited and when our ancestors lived in conditions that were, in some ways, only slightly above those of animals today. And since that time, human history has seen the gradual emergence of the family, the tribe, the race, the city-state and, finally, the nation-state as progressively more complex forms of social organization.

Particularly important was the transition from the tribe to the higher, more complex units, because this transition seems to have depended on the acquisition of a basic new social skill, namely the ability for so-called one–many relationships. It appears that all forms of social organization on the tribal or pretribal level are based exclusively on one–one relationships in which each member of the group knows personally every other member of the group. It was only when individuals were able to relate, not only to other individuals but also to abstract groups of individuals, that such units as the nation-state became possible.[33] For example, at the apogee of the Islamic nation-state, it was possible for a Muslim to travel in perfect safety from, say, southern Spain, through the whole Mediterranean Basin, to India, encountering only personal strangers but being accepted everywhere as a fellow Muslim.

The third basic stage in collective human growth, that of secondary integration, corresponds to maturity or adulthood in the life of the individual. The individual, having successfully forged his identity and fully developed his powers, now consciously and deliberately seeks self-integration through a new synthesis, a synthesis based on the analytical and critical distinctions he has made and the differentiated capacities he has developed. Thus, to achieve its collective maturity, humankind must move forward towards an entirely new synthesis that builds upon the differentiated identity and collective self-awareness generated by science. It must also acquire yet a further social skill, namely the capacity for many–many relationships – for harmonious relationships between groups, peoples, and nations.

According to the organismic theory of history, we have not yet achieved our collective maturity. Rather, we are currently in the throes of late adolescence, with its turbulence, its *Sturm und Drang*. Secondary integration, synthesis, unity in diversity – this is the goal towards which we are moving. It is the consummation of human social evolution, the ultimate stable configuration of human society, the adulthood of the human race on this planet:

> The long ages of infancy and childhood, through which the human race had to pass, have receded into the background. Humanity is now experiencing the commotions invariably associated with the most turbulent stage of its evolution, the stage of adolescence, when the impetuosity of youth and its vehemence reach their climax, and must gradually be superseded by the calmness, the wisdom, and the maturity that characterize the stage of manhood. Then will the human race reach that stature of ripeness which will enable it to acquire all the powers and capacities upon which its ultimate development must depend.[34]

The stability represented by maturity does not imply that no further change or progression occurs. It means rather that future change takes place under different conditions and in different ways than was previously the case:

> The emergence of a world community, the consciousness of world citizenship, the founding of a world civilization and culture . . . should, by their very nature, be regarded, as far as this planetary life is concerned, as the furthermost limits in the organization of human society, though man, as an individual, will, nay must indeed as a result of such a consummation, continue indefinitely to progress and develop.

That mystic, all-pervasive, yet indefinable change, which we associate with the stage of maturity inevitable in the life of the individual . . . must . . . have its counterpart in the evolution of the organization of human society. A similar stage must sooner or later be attained in the collective life of mankind, producing an even more striking phenomenon in world relations, and endowing the whole human race with such potentialities of well-being as shall provide, throughout the succeeding ages, the chief incentive required for the eventual fulfillment of its high destiny.[35]

Just as maturity for the individual involves the emergence of a new kind of wholeness that integrates the multifarious abilities developed during adolescence, so the maturity of humankind can be achieved, not by a regression to the undifferentiated wholeness of mythmaking or traditional mysticism, but only through a new and creative wholeness that crucially depends on the analytic, differentiated scientific knowledge resulting from the systematic application of the model-building paradigm.

Thus, from the viewpoint of the organismic theory, the crucial mistake of modern civilization was in taking an intermediate stage of social evolution as the final and ultimate configuration, in taking secular, scientific man as prototypical human, as human being *par excellence*. In taking this partially complete product as the final and ultimate one we have denied our basic need for wholeness and for a relationship with what is ultimate in invisible reality — whence the alienation that is so characteristic of modern life.

Of course, this was a very easy mistake to make. Model building was quickly successful once systematically applied, even by a small segment of society. The new paradigm was so obviously superior that it was natural to suppose that science alone could be the answer to our quest for meaning in existence. But if we view model building not as the final paradigm, but rather as a necessary (and irreversible) inter-mediate step towards the final paradigm, then we see that it is neither necessary nor helpful to abandon model building in order to achieve the wholeness we seek. We do not have to destroy, only to move forward towards the new, creative synthesis of secondary integration.

It is possible to characterize the nature of secondary integration in a variety of more or less equivalent ways: unity in diversity, the unity of religion and science, the equality of men and women, the unity of races, the political unity of nations. Each of these 'unity of opposites' is like a projection or representation of secondary integration.

It is not the purpose of the present essay to develop in further detail the organismic theory of history, to attempt a more critical analysis of the nature of secondary integration, or to arrive at a precise and satisfactory understanding of all that is involved in the transition from collective adolescence to collective maturity.[36] But, to the degree that the organismic theory is correct, we can already see that we must strenuously resist the temptation to abandon model building in our desperate desire to achieve wholeness. Rather, we must bring to bear on the present configuration of the collective organism that is mankind the full force of our most mature thought and our most constructive efforts in a sincere and concerted attempt to give birth to the new age of a dynamic and progressive, yet stable and peaceful, society.

The characteristic features of the stage of secondary integration can, for the most part, only be achieved as a result of conscious and deliberate effort. The transition from childhood to adolescence results from processes over which the individual has little control. Similarly, the chief features of the present age in human history are not the result of much conscious planning on our part. But the coming age of universal civilization and culture can only be accomplished through intelligent and sustained effort.

Both the promise of success and the realization of the effort necessary to achieve this success are summed up in the following words of 'Abdu'l-Bahá with which we end our essay:

A few, unaware of the power latent in human endeavor, consider this matter as highly impracticable, nay even beyond the scope of man's utmost efforts. Such is not the case, however. On the contrary, thanks to the unfailing grace of God, the loving-kindness of His favored ones, the unrivaled endeavors of wise and capable souls, and the thoughts and ideas of the peerless leaders of this age, nothing whatsoever can be regarded as unattainable. Endeavor, ceaseless endeavor, is required. Nothing short of an indomitable determination can possibly achieve it. Many a cause which past ages have regarded as purely visionary, yet in this day has become most easy and practicable. Why should this most great and lofty Cause – the day-star of the firmament of true civilization and the cause of the glory, the advancement, the well-being and the success of all humanity – be regarded as impossible of achievement? Surely the day will come when its beauteous light shall shed illumination upon the assemblage of man.[37]

3

From Metaphysics to Logic

A Modern Formulation of Avicenna's Cosmological Proof of God's Existence

Determining the exact status of purported proofs of God's existence is not an easy matter. Inductive proofs (which involve generalizing from particular instances to abstract principles) are convincing only if one is truly convinced that the inductive leap is logically justified, a question about which fair-minded assayers of the argument may differ.

Deductive proofs have the advantage that the logical principles used can be made explicit (in fact, they can be formalized), thereby avoiding purely logical disputes among all who accept the system of logic in question. But deductive proofs must proceed from premises (hypotheses, assumptions) whose validity can be open to question. This is particularly so when the hypotheses involve abstract philosophical notions, which are often inherently imprecise. Moreover, the logical structure of deductive proofs is such that strong conclusions require strong premises. Since the existence of God is a very strong conclusion, the assumptions from which it is deduced

This essay was first presented at the Autumn 1989 Logic Colloquium at Université Laval, Québec, under the title 'La démonstration de l'existence de Dieu chez Avicenne'. This is its first publication.

must likewise be strong, and one is frequently left with the feeling that the proof has only established the truism that if necessarily God exists, then God exists.

A case in point is Aristotle's 'cosmological' or 'first cause' argument.[1] Starting from observations about causal relations in physical processes, Aristotle reasons, in effect, that there must be an uncaused cause (an unmoved mover), for otherwise we would be faced with an infinite regression of causes, a configuration held to be logically impossible.[2]

A modern criticism of Aristotle's argument might start by pointing out that we now know from results in mathematics and logic that there is nothing inherently contradictory about the notion of infinity or even about the notion of a discretely and totally ordered infinite set having no minimal element (which is just a precise definition of an infinite regression).[3] Of course, an infinite regression of *causes* is admittedly more difficult to imagine. However, 'cause' is a good example of a notion that can be analysed in a number of logically different ways, some compatible with the notion of an infinite regress and others not.[4]

Another kind of difficulty with a proof like Aristotle's lies in assessing exactly what the proof proves. Granting that we have proved the existence of an unmoved mover or an uncaused cause, to what extent are we justified in calling such an entity God or the Creator, more especially if we attribute consciousness and deliberateness of purpose to God? Maybe the universe itself is the uncaused cause of everything within it, without the universe being a conscious, intelligent, or willing agent.

At the very least, Aristotle's argument is subject to the following dilemma: If, on the one hand, we hold as a logical principle that every existing entity is caused (by some agent other than itself), then an uncaused (i.e. self-caused) cause violates this very principle and cannot, therefore, exist. On the other hand, if we admit the possibility that some entities may exist without a cause, then on what basis do we attribute primacy of generation to some first cause? Perhaps there are many instances of uncaused entities, even among observable objects. (In this regard we should remind ourselves that a causal link or connection between two entities is always *inferred* from observation, never directly observed.)

Over the years, many philosophers have identified such weaknesses in proofs of God's existence, but few have equalled Avicenna (Ibn Sina) in the vigor and clarity with which he has undertaken to repair the defects and remove the lacunae in these arguments. In particular, with regard to Aristotle's first-cause proof, Avicenna devised a brilliantly-conceived variant based on several novel ideas and insights that show how clearly Avicenna understood the weaknesses in Aristotle's argument. The present study proposes, first, to examine in some detail Avicenna's variant of Aristotle's cosmological proof of God's existence and then, in turn, to assess Avicenna's argument in the light of certain principles of modern logic. Finally, we will use our analysis of Avicenna's argument as a basis for recasting it in a more logical and less metaphysical mold.

Avicenna's Proof

In a certain sense, Avicenna's proof really begins where Aristotle's leaves off.[5] To be more precise, Aristotle's proof seeks to establish the existence of an uncaused, prime cause, and the main portion of his argument is applied to that end. Aristotle appears to take for granted that, when once the existence of his unmoved mover is established, nothing further is needed.[6] Thus, Aristotle never directly addresses the dilemma described in our introduction above, while Avicenna deals with that dilemma at the very outset of his proof.

Avicenna begins with an analysis of the notion of causality. After considering a number of different instances of causal laws, he arrives at two basic categories of existence: (a) Those existing entities (beings) whose existence is caused by some other entity; (b) those existing entities that are uncaused, i.e. that are sufficient for their own existence. Avicenna then undertakes a critical analysis of entities of the second category (b).

A self-sufficient entity cannot 'have a cause', for if it did, then it would exist by virtue of that cause rather than by virtue of itself alone.[7] Important for Avicenna's argument (as will be presently clear) is that he excludes not only external agents as causes of an entity in category (b) but also causes that are internal to the entity. For if a self-sufficient entity had a cause internal to it, it would owe its existence to some *part* of itself rather than to itself.

Let us amplify this a bit. An entity can be composed of other entities, or it can be uncomposed or simple (irreducible). Avicenna held that a composite entity cannot be self-sufficient, for a composite entity exists by virtue of the components which make it up, rather than by virtue of itself as a whole, distinct from its parts.[8] It follows, asserted Avicenna, that any entity in category (b) must be simple and incorporeal (for otherwise it would be composed of physical parts). It must, he claimed, be a pure, undefinable essence, an essence whose existence is identical to it.[9] Finally, Avicenna argues that there can be at most one such entity. For if there were two, he says, then at least one of the entities would be composite, possessing both that by which the entities were similar (their each being self-sufficient) and that by which they were different.[10]

Thus, by his analysis of the notion of self-sufficiency, Avicenna seeks to respond in advance to one of the most telling criticisms of Aristotle's first-cause proof, namely that Aristotle fails to justify the identification of his prime mover with God. In particular, Avicenna's exclusion of composite entities from category (b) completely sidesteps any arguments to the effect that some physical entities may be uncaused.[11] Of course, the logical and/or philosophical correctness of Avicenna's analysis is one of those questions about which equally fair-minded philosophers may differ.

Therefore, to sum up this initial portion of Avicenna's cosmological proof: Avicenna holds that there are four logical categories of existence: caused and uncaused (self-caused), composite and simple. In particular, uncaused entities may exist, but are always simple. He further argues, in an analysis of the notion of self-sufficiency, that there is at most one uncaused entity which, if it exists, may be reasonably identified with God (the Creator). [12] Avicenna now turns to the remaining task of establishing the existence of such an entity by a proof that avoids the pitfalls of the infinite regression argument of Aristotle.

Let us consider, says Avicenna, the collection C of all caused entities that exist at the present moment. C may be finite or infinite; it doesn't really matter which is the case. (It is even conceivable that, at some moments, there are an infinite number of things in existence while, at other times, only a finite number.) But, since we can observe that various physical objects exist at any given moment, the collection C is not empty. C is thus a *composite entity* and cannot, therefore, be self-sufficient. It must have an existing entity E (different from the collection C itself) as a cause. Moreover, E must be outside of C, for suppose not. Then E would be both a member of the collection C and the cause of C. E would therefore be its own cause, i.e. self-sufficient, contradicting its status as a member of C (and thereby caused). Hence, the cause E of C must be outside of C, and thus uncaused. But, as our previous arguments have shown, there can only be one uncaused entity, namely God. Thus, E = God, i.e. God exists. [13]

Evaluating the Proof

A proof like Avicenna's can be evaluated in several different ways. We can, for instance, consider the philosophical plausibility of its hypotheses. Likewise, we can attempt to evaluate the validity of the logical principles used in the course of the proof. Finally, we can assess the overall believability of the proof: is it intuitively or psychologically convincing?

On the whole, Avicenna's proof carries a fairly high degree of intuitive conviction. The premises of the proof appear to take careful account of the various logically possible categories of existence (the caused, uncaused, composite, and simple). Moreover, there is no immediately apparent logical flaw in the reasoning used. Nonetheless, there are logical problems with Avicenna's proof, particularly in connection with his strategy of avoiding the Aristotelian infinite regression by considering the collection C of all caused entities to be a single composite entity. [14]

In fact, the kinds of logical problems raised by Avicenna's proof have received systematic treatment in logic and mathematics only in the late nineteenth and early twentieth centuries, beginning with the brilliant work of Georg Cantor, the 'father' of modern set theory (and in particular of the modern theory of infinite sets). [15] Avicenna's idea of considering a collection of entities to be a single composite entity can therefore be seen as an interesting (and in some ways exciting) anticipation of developments in modern logic, and the problems it raises have still not been resolved in any generally accepted and satisfactory manner, though considerable clarification and understanding have been achieved.

Let us begin our analysis of these logical problems by raising the following question: does every class (collection) of entities constitute a single composite entity? There is no clear and immediate answer to this question. Cats are entities, but the collection of all cats is not a cat. Can the class of all cats be reasonably considered an entity? The answer clearly depends on how liberally we construe the meaning of the word 'entity'. The class of all cats is certainly not an organism in the accepted sense of this latter term. But does it really matter whether we consider such collections as the class of cats to be entities?

The answer is 'yes', in some contexts it matters quite a bit.

In particular, it is crucial to the logic of Avicenna's proof that the class C of all caused entities be itself an entity, for if not, then Avicenna is not logically justified in applying to C all of his previous analysis concerning uncaused and composite entities, and it is on the basis of this analysis that he seeks to justify his conclusion that the cause E of C must be outside of C and thus uncaused.

In other words, if we assume the principle that all composite *entities* are caused, then a composite (class) that is not an entity could still be uncaused. If, therefore, the class C of all caused entities is not an entity, then we can logically hold C to be uncaused even though each member of the class C is caused. [16]

Even more generally, we can observe that whatever is true of every member of a collection is not necessarily true of a collection. For example, at the present moment my body is an entity (organism) which can be logically viewed as the collection of all those cells that currently make it up. Moreover, no cell of my body can walk around on its own power. But my body as a whole can walk around on its own power. Thus, what is true of each cell of my body is not true of the body as a(n) whole (entity). In the same way, to say that every member of a collection is caused does not necessarily mean (i.e. logically imply) that the collection itself is caused.

Thus, it is of pivotal importance to Avicenna's argument that the class C of all caused entities be itself an entity, for that is the only thing which logically justifies the application of Avicenna's various principles of causality and existence to C.

Suppose, then, that we try to overcome this difficulty by adding a further assumption to Avicenna's argument, namely, that the class C of all caused entities is an entity. C is now a composite *entity* and thus caused, and the rest of the argument continues as before, leading logically to the conclusion that God exists. But, unfortunately, new difficulties now arise.

C is now both a caused entity itself and the collection of *all* caused entities. Thus C is a *member of itself*, a seemingly absurd situation whose absurdity can be seen in the following way: It seems clear that we cannot actually form a collection of objects until every object to be included in the collection has already been formed (e.g. my body at the present moment cannot exist *before* each cell that presently makes it up exists). Thus, the collection C cannot exist until every caused entity exists. But if the collection C is itself a caused entity, then C cannot exist until C exists, until C exists . . . an infinite regression (or a circularity)! Thus, the validity of Avicenna's argument would appear to hinge on the legitimacy of an infinite regression or a circularity, and one that seems quite unacceptable and absurd. For future reference, we will call this logical problem the 'self-membership paradox'. [17]

Let us make another attempt to repair the logical defects in Avicenna's proof by redefining C to be the collection (composite) of all caused entities *except C itself*. Then, regarding C as an entity, we can correctly apply Avicenna's principles concerning causation and composite entities to C, concluding that C must have a cause E that is outside of C. Of course, C itself is an entity outside of C, and is thus a possible cause of C. However, C is a composite entity and cannot, therefore, be its own cause, according to Avicenna's principle that no composite entities are self-caused. Thus, the cause E of C is outside of C *and* different from C. Thus E is the uncaused cause of C.

This latter argument would appear to have solved our difficulties, but has it? Not quite, for crucial to this last solution is the principle that the class C of all caused entities different from the class C itself be an entity (composite, of course). But then what about the further collection C^*, which consists of C augmented with C itself, and the collection C^{**}, which is obtained from C by the addition both of C and the uncaused cause E of C as further members? C^* has only one more member than C, namely C itself, and C^{**} has only two more members, namely C and E. Are C^* and C^{**} entities?

It is very difficult to think of any logically coherent criterion of entityship that would consider C as an entity but deny entityhood to C^* and to C^{**}. Yet, this is what must be the case if we are to avoid further paradox. Indeed, since the simple entity E and the composite entity C are the only two existing entities not included in the collection C itself, C^* is the class of all caused entities (E, the only entity outside C^*, is uncaused), and C^{**} is the class of all entities in existence (which we will symbolize also by V). If C^* is an entity, it is composite and therefore caused. Hence, it is a member of itself, and we are again confronted with the self-membership paradox. Likewise, if $V = C^{**}$ is an entity, it also is a member of itself since V is the class of all entities in existence. (This latter form of the self-membership paradox is sometimes referred to as the 'universal set paradox'.)

Let us suppose, then, that C is an entity but that C^* and C^{**} are not. It now follows logically and consistently that C has an uncaused cause E, i.e. that God exists (provided Avicenna's various other principles regarding causality and the composition of entities are correct).

But how reasonable are these entityship assumptions about C, C^* and C^{**}? Not very. Why should the class C of all caused entities different from C constitute an entity, while the addition of only one or two new objects to C destroy that property? Of course, we have not given any criterion of entityship, and that is the very problem. We have simply assumed, in a totally arbitrary fashion, that C is an entity and that C^* and C^{**} are not, because those are the assumptions which allow us to avoid contradiction and to conclude that God exists. But unless we have some natural, intuitive notion of composite entities that accepts C and rejects C^* and C^{**}, then the word 'entity' is a meaningless label.[18] We might just as well assume that God exists and forget logic altogether.

Thus, the argument for God's existence is, in its present form, logically correct (the conclusion follows logically from the hypotheses) but intuitively and psychologically unconvincing because of the gratuitous nature of some of the assumptions regarding entities upon which the argument is based.

Putting Things Right

Our solution to the entityship problems in Avicenna's proof was unsatisfactory because it was *ad hoc*. The proof, as it now stands, is a compromise between the somewhat subtle logical problems arising from the method Avicenna was pioneering, on the one hand, and the various metaphysical considerations underlying Avicenna's principles of causation and entityship, on the other. But, as we shall soon see, judicious appeal to a few principles of the modern logic of set theory will allow us to reconstruct Avicenna's proof in a way that places its essential burden on the following *contingency principle*: No phenomenon all of whose components are caused can itself be uncaused. [19]

We begin with a few definitions. We let V stand for the *universe*, i.e. the collection of all entities in existence. Thus, every existent being (force, entity, relation, object, form, idea) is an element of (a member of) the collection V. [20] By a *phenomenon* we understand any collection (class, composite) of entities (called the *components*, the *elements* or the *members* of the phenomenon). Thus, a phenomenon is a portion of (a subclass of) the universe V. More generally, we can say that a phenomenon A is a *subclass* (*subphenomenon*) of the phenomenon B if every entity in the class A is also in the class B. A subclass A of B is *proper* if $A \neq B$. Associated with any finite number of entities $x_1, x_2, \ldots, x_n$ is the phenomenon $\{x_1, x_2, \ldots, x_n\}$ comprised of exactly those entities. Thus, every entity x

determines a one-element phenomenon {x}, and every component x of a composite phenomenon A determines a one element
subphenomenon {x} of A.

As we have already noted in the preceding section, some
entities are themselves classes, i.e. collections of other entities.
If a phenomenon (class) A is an entity, then we call A a *set*; if
not, then it is a *megaphenomenon* (or a *big class*). An entity that is
not a set is said to be *simple* (or *atomic*) and is called an
individual (or an *atom*). Thus, to sum up, our ontology
comprises the following distinct categories: (1) simple entities;
(2) composite entities (sets); (3) megaphenomena (big classes).[21]

We now make the explicit assumption that the universe V is
well-founded.[22] Among other things, this means that no class
can be an element of itself and that the membership structure
of every class has finite depth. On the practical level, this latter
means that every nonempty class is built up by starting with
simple entities or sets and forming classes of classes of . . .
entities, with only a finite number of iterations of class
formation.

We now come to one of the main differences between the
present approach and that of Avicenna. Whereas Avicenna
appears to have considered causality as a relationship between
entities only (i.e. one entity is or is not the cause of another
entity), we affirm that it is more properly viewed as a
relationship between phenomena and entities (one phenomenon or entity is or is not the cause of another phenomenon or
entity). Indeed, in science most of the phenomena studied are
composites of several entities. Similarly, the cause of a given
phenomenon may also be a composite of several entities. As
will be seen, this modification allows us to avoid almost
entirely any problems of entityship because, for the most part,
it will not matter whether or not a phenomenon is an entity.

To facilitate our discussion, we make one further termino-
logical convention concerning the use of the term 'phenomenon'. As it now stands, a phenomenon P is a composite of

entities, whether P is an entity (set) or not. Thus, the only ontological category that is not a phenomenon is a simple (noncomposite) entity. We hereby extend the definition of 'phenomenon' to cover the case of simple entities. If necessary to avoid ambiguity (but not otherwise), we will speak of simple (or noncomposite) phenomena and composite phenomena. Thus, a composite phenomenon may or may not be an entity, but a simple phenomenon is always an entity. With this terminology, causality is a relationship between (simple and composite) phenomena.

Where A and B are two phenomena, if A causes B we say that A is a *cause* (of B) and B an *effect* (of A), and we express this relationship symbolically by writing $A \rightarrow B$. If $A \rightarrow A$ we say that A is *uncaused* (or *self-caused*). If B has a cause A different from B, we say that B is *other-caused* (or, simply, *caused*). Basic to all scientific activity is the *causality principle*: Every existing phenomenon is either self-caused or other-caused (but not both). We also assume the *transitivity principle*: If $A \rightarrow B \rightarrow C$, then $A \rightarrow C$.

The causality and transitivity principles together imply that there can be no circular causal chains among distinct phenomena, for suppose there is such a chain $A_1 \rightarrow A_2 \rightarrow \ldots \rightarrow A_n \rightarrow A_1$. Then, by the transitivity principle, each element A_j of the chain is self-caused, i.e. $A_j \rightarrow A_j$, and thus, by the causality principle, no element of the chain can be other-caused. But every element of the chain is caused by every other element in the chain. Hence, the elements of the chain are all the same, i.e. $A_1 = A_2 = \ldots = A_n$.[23]

The principles of causality and transitivity have been purposely chosen to be minimal in what they suppose about the causality relation. This is one of the reasons why we have not attempted to define causality in set-theoretic terms. When once we have completed our argument, we will indicate how such a set-theoretical definition of causality can be accomplished (at that point it will be obvious to the reader in any case).

However, as things now stand our treatment of causality is defective in that there is no principle linking the causality relation between A and B with the set-theoretic structure of A and/or B. We remedy this defect with the *potency principle*: If A causes the composite phenomenon B, then A → C where C is any subphenomenon or any component of B.[24]

Of course, we cannot actually see causal links themselves and verify through direct observation that the above (or any other) principles are true. But we infer the existence of causal links from our observations of concrete phenomena. We then make mental models of the world based on the assumption that these links exist objectively and that they satisfy our principles (among others). We then proceed to interact with reality on the basis of these models and these assumptions, making predictions about what will happen under certain circumstances and expecting these predictions to come true. And, very frequently, these predictions do come true (and when they don't we can often discover, in retrospect, previously hidden reasons (causes) to explain why our initial prediction failed).

Nevertheless, it is logically possible that all this is nothing but a monstrous illusion. It could be that there are no causal links at all and that the many regularities we observe (or feel we observe) have been, until now, just the result of a series of highly unusual coincidences. Perhaps we will wake up tomorrow to an experience of utter chaos in which nothing behaves as we have come to expect. However, this logical possibility seems *much* less likely than does the meta-hypothesis that objective causal links exist. Our unfulfilled predictions and expectations can be easily explained as errors in our models (which we know to be approximate in any case) rather than as evidence for the absence of causal links. We therefore hypothesize that there are objective causal links between the various phenomena in existence. To do otherwise would be grossly illogical and unscientific.

Since we are mainly concerned here with proving the existence of God, we need to make as precise as possible what we mean by 'God'. Let us take the concept of God on which Avicenna's proof is based, i.e. God as Creator. In scientific terms, this means that God is an existing force or entity that is responsible for (the cause of) the entire universe itself (and thus, by the potency principle, the ultimate cause of every force or entity in the universe). God is thus defined as a *global or universal cause*.

More generally, science has been led to postulate the existence of a certain number of unseen forces as causes of various observed phenomena. For example, the unseen force of gravity is the cause of the behavior of unsupported objects in the presence of a large mass such as the earth; electromagnetic force is the unseen cause of the visible light produced by a glass-enclosed metal filament in a closed electrical circuit. The idea that there could be some single, unseen force ultimately responsible for all forces or entities in the universe is scientifically coherent; indeed, it is a very natural hypothesis.[25] Moreover, if such a force or entity exists it is unique, for suppose there were two such force-entities. Then each would be both self-caused and other-caused, contradicting the causality principle.

The task now before us is to prove, on the basis of our causality principles, that a (necessarily unique) universal cause exists. Our discussion of these matters will be facilitated if we pause here to establish some further terminology. When A → B → C, we say that B is *intermediate* between A and C. If A causes B and there is no intermediate between them, we say that A *directly* causes B. A *causal chain* is a sequence of cause and effect relations. A causal chain K is *explicit* if every link is direct. A (nonempty) causal chain in which every cause has a *new* (i.e. not previously appearing in the chain) effect is said to be *infinite ascending*; if every effect has a new cause, then K is *infinite descending* or an *infinite regression*. A non infinite-ascending causal chain is *upward finite* and a non infinite-descending causal chain is *downward finite*. A *finite* chain is one that is both upward and downward finite. A finite chain has both a first and a last element which we call the *initial cause* and the *final effect*, respectively, of the chain. Finally, a finite chain is *complete* if its initial cause is uncaused.

We look first at Aristotle's proof of the existence of an uncaused cause. His proof is based on the assumption (principle) that there can be no infinite descending causal chain, for suppose this principle holds. Then, starting with any given effect A, we can construct a descending causal chain $B_n \to \ldots \to B_1 \to A$, whose final effect is A, by adding causes of causes, etc., each newly-added cause being new to the chain. Such a chain will be infinite descending unless we eventually arrive at an effect B_n for which no new cause exists. By the principle of causality, B_n is thus a self-caused cause. (Notice that if the initial effect is uncaused, the $B_n = A$.)

In other words, if Aristotle's principle is true, then every effect A in the universe is the final effect of an uncaused cause B that is the initial cause of a complete finite chain from B to A. The plausibility of Aristotle's proof is more or less directly proportional to the plausibility of the principle that no infinite descending causal chain can exist. We leave the reader to decide for himself what he feels that latter plausibility to be.

But, questions of plausibility aside, even if we grant Aristotle's principle and thereby accept that an uncaused cause exists, we still have not established the existence of God as defined above, i.e. as a universal cause. Nor have we proved the uniqueness of an uncaused cause. Indeed, if Aristotle's principle is true, there could still be any number of different uncaused causes occurring as initial causes of different complete finite chains. There could even be two different complete finite chains whose final effects are the same (in other words, different causal chains starting at different uncaused causes could still yield the same final effect).[26]

In fact these observations are nothing but a somewhat more precise rendering of the weaknesses in Aristotle's proof that Avicenna clearly saw and endeavored to remove. So, let us see how Avicenna's proof fares in the present context. We begin by postulating the previously mentioned *contingency principle*: If every component (element) of a composite phenomenon A is caused, then A itself is caused.[27]

Lemma 1. If A is an uncaused phenomenon, then A is simple (noncomposite).

Proof. Let A be some uncaused phenomenon, A → A. Suppose A were composite. Then, by the contingency principle, some component B of A is uncaused (otherwise, every component of A is caused, implying that A is caused). Now, since A → A, A causes every component of A by the potency principle. Thus, A → B (since B is a component of A). But B is uncaused. Thus, A = B by the causality principle, implying that A is a component (member) of A. But this latter is impossible by the well-foundedness principle. Hence, our supposition that A be composite is false, and A is simple as claimed. ■

Because we assume the contingency principle, the statement just proved is equivalent to the following:

Corollary of Lemma 1. Every composite phenomenon A is caused. ■

We now prove a further lemma:

Lemma 2. If A is a composite phenomenon and if A has some uncaused component B, then B → A.

Proof. Since A is composite, it is caused (by the above Corollary). Let K be a cause of A, K → A. By the potency principle, K causes every component of A; hence, K causes B. But B is uncaused. Thus, by the causality principle, K = B and B → A as claimed. ■

We prove one last lemma before attacking the main proof.

Lemma 3. There is no more than one uncaused phenomenon E and, if it exists, E is a (simple) entity in the universe V.

Proof. By Lemma 1 any uncaused phenomenon is simple and therefore an entity. Suppose there were two uncaused entities, E_1 and E_2. We form the composite phenomenon $P = \{E_1, E_2\}$. E_1 is therefore an uncaused component of P. Hence, by Lemma 2, $E_1 \rightarrow P$. But E_2 is also a component of P. Thus, by the potency principle, $E_1 \rightarrow E_2$. However, E_2 (as well as E_1, of course) is uncaused. Hence, by the causality principle, $E_1 = E_2$. ■

We now have:

Theorem 1. There exists a (unique) universal cause E.

Proof. Let C be the phenomenon consisting of all caused entities in the universe V. Clearly, C is composite. Thus, by the Corollary to Lemma 1, C is caused by some phenomenon A, $A \rightarrow C$. There are two cases to consider. Suppose, first, that A is simple. Thus, A is an entity. If A were caused, then A would be an entity in C. Thus, by the potency principle, we would have $A \rightarrow A$, i.e. A is uncaused! This contradiction means that A is not a component of C, i.e. A is uncaused and therefore, an uncaused cause of C.

Suppose, now, that A is composite. Suppose, further, that all components of A are caused. Then A is a subphenomenon (subclass) of C. Since $A \rightarrow C$, A is a cause of every subphenomenon of C, by the potency principle. But A is a subphenomenon of C. Thus, $A \rightarrow A$, i.e. A is uncaused. But A is composite and therefore caused (Lemma 1). This contradiction means that A cannot consist only of caused entities (in other words, A cannot be a subclass of C). Thus, A must contain at least one uncaused entity B. By Lemma 2, we thus conclude that $B \rightarrow A$. Since we already have $A \rightarrow C$, an application of the transitivity principle[28] yields that $B \rightarrow C$, i.e. *B is an uncaused cause of the class C of all caused entities.*

We have now shown that, in all cases, there is an uncaused entity that is the cause of the class C of all caused entities. By Lemma 3, there is only one uncaused entity in the whole universe V. Thus, we now give the name E to this one uncaused entity. We have thus established that $E \rightarrow C$. To complete our proof, we need to show that E is, in fact, a universal cause, i.e. the cause of all phenomena in the universe.

Since $E \rightarrow C$, the potency principle tells us immediately that E is the cause of every caused entity in the universe and, indeed, of any composite phenomenon all of whose components are caused. Suppose P is a composite phenomenon one of whose components is the uncaused entity E. Then, by Lemma 2, we immediately have $E \rightarrow P$. Thus, E is a universal cause, the cause of every phenomenon in the universe (including the megaphenomenon V that is the universe itself). ■

Our proof is now complete. We close this section of the essay with a description of a set-theoretical definition of the causality relation. Let V be the collection N of positive integers $\{1,2,3, \ . \ . \ .\}$ to which we adjoin one new object E. All objects in V are simple entities (phenomena), while collections (classes) of objects in V are composite phenomena. The causality relation $A \rightarrow B$ between two phenomena A and B is defined as follows:

(1) If $A = E$, then $A \rightarrow B$, for any B whatever. If B is the entity E, or a class having E as a member, then E is the only cause of B.

(2) If B is an entity $n \neq E$, then $A \rightarrow B$ if A is a class having n as a member (component). If B is a class of which E is not a member, then $A \rightarrow B$ whenever B is a proper subclass of A.

It is easy to check that the four principles of causation are satisfied by this definition. (a) Since no class is a proper subclass or a component of itself, E is the only self-caused phenomenon, and every other phenomenon has E as a cause (by (1)). Thus, the causality principle is satisfied. (b) The transitivity principle is satisfied since E is a universal cause and since the relationship of class containment is transitive (A a subclass of B, and B a subclass of C, implies that A is a subclass of C). (c) The potency principle is satisfied by the same reasons as in (b). (d) The contingency principle is satisfied since every class A not containing E is (other-) caused (by E).

As a matter of fact, the four causation principles are satisfied if we define causality only by the clause (1) above, but the causality relation is not so interesting in that case. Indeed, there are some very trivial models of the four causation principles (just let the universe V consist of E with E → E), but such models are not very instructive.

The model based on clauses (1) and (2) above has one very interesting property: it does *not* satisfy Aristotle's principle forbidding infinite regressions of causes! Start with any class not containing E, say the empty class o. Then the system of proper supersets of o not containing E has an infinite descending causal chain with final effect o (just extend by adjoining the natural numbers one by one, obtaining the chain . . . →
. . . → $\{1, 2, \ldots, n\}$ → . . . → $\{1, 2\}$ → $\{1\}$ → o). This shows that Aristotle's method and Avicenna's method are logically independent of each other. It also shows that we do not need to appeal to Aristotle's principle in order to prove the existence of an uncaused cause. *In fact, we have proved the existence of a universal uncaused cause without explicitly assuming the existence of an uncaused cause or even of a noncomposite entity.*

Evaluation and Conclusion

We have succeeded in replacing most of Avicenna's metaphysical assumptions by purely logical principles that represent truths of the causality relation as modeled by modern scientific practice. On the basis of these principles, we have proved (without any appeal to modal logic whatever) that a universal uncaused cause exists. Particularly striking is the fact just mentioned, that we prove the existence of a simple, uncaused cause without even assuming explicitly the existence of a simple entity.

If, as seems quite justified, we take the causality, transitivity, and potency principles as pragmatically verified by modern scientific practice (or, what is more or less the same thing, as principles inherent in the very logic of causality itself), it remains to assess the contingency principle upon which the burden of our proof rests. [29]

The main difficulty in confronting this task is that our experience of observable, physical reality is one in which we seem to encounter only caused entities. But, the (sub)universe (or megaphenomenon) of *all* physical (and, presumably, caused) entities engulfs and surpasses us. We have no direct way of verifying what laws or principles such a vast system might satisfy.

We might begin by reasoning as follows: Since (as we assume) caused entities do not have the power to come into existence on their own, the physical universe would not exist if it were not caused. Yet, there is the possibility that the physical universe is uncaused but has always existed. [30] Perhaps it never *had* to come into existence. Under these hypotheses, the physical universe would be a closed and isolated system, a megaphenomenon with no beginning and no end, no first principle and no final goal. What would we expect such a system to be like? How could we know if this were so?

The best we can do in trying to answer this question is to reason by analogy with particular phenomena we have studied and understood to some degree. Though no physical system we have studied is a perfectly closed and isolated system, we have nevertheless studied a number of systems that are relatively closed and isolated. Their most pervasive feature is: *they all degenerate*. This is the well-known second law of thermodynamics, that, in a closed system, entropy (disorder) increases until a state of maximum entropy (total disorder) is attained.

Thus, if the physical universe is uncaused, it is an isolated system that has always existed. It should, therefore, be in a state of maximum disorder and chaos. However, we observe that there is a very pervasive and refined order in the universe.

This fact renders very implausible the thesis that the physical universe is uncaused, and therefore very plausible the thesis that it is caused.

How valid is our extension of contingency properties of observed phenomena to contingency properties of the physical universe and then to the universe of all caused entities? This, once again, is a matter about which fair-minded people may well disagree. In any case, the method of extending results and principles deriving from our observations of limited phenomena to the physical universe as a whole is the basis of all current scientific cosmological theories. There is no *prima facie* reason why it should not be valid in the present context.[31]

4

A Logical Solution to the Problem of Evil

In this essay we will discuss the philosophical problem known as the 'problem of evil'. The classic form of this problem runs something as follows: If there is a God, then he cannot be both omnipotent and good. For, since there is evil in the world, God, if he be all-powerful, is responsible for this evil (since he could prevent it if he chose) and is thus himself evil.

The problem is a real one, for the choice which seems to be imposed by the above argument is hard indeed. If God really is not all-powerful, but is good, then what is the limit of his power? Precisely, evil and his inability to conquer it. Certainly, a good God must wish to overcome evil, and since he evidently has not, it follows that it is because he has not been able to do so. Thus, evil and its force would seem to be more powerful than such a God, and he ceases to be any sort of God at all. He is, at best, a sort of ally with us (or some of us) in the struggle against evil.

This essay is reprinted with permission from *Zygon*, vol. 9, no. 3 (September 1974), 245–55.
© 1974 by The University of Chicago.

On the other hand, an all-powerful but evil God is equally unsavory to contemplate.

Logically speaking, there is one simple way out of the dilemma: Deny the existence of evil. If there is no evil, then God can logically be held to be both good and all-powerful. Among those thinkers who have squarely faced the problem (and there may not be too many), some, such as Leibniz, seem to have chosen this way out.

But if the above is logically satisfying, it is certainly not, at first glance in any case, emotionally and morally satisfying. Our moral repugnance (or at least the moral repugnance of a certain large proportion of the world's population) at such atrocities as death camps, genocide, homicide, war, persecution, etc. makes it difficult for us to believe that evil does not exist. If there is no evil, then there is certainly an abundance of suffering and injustice. And if suffering, or at least injustice, is not evil, then are we not simply playing with words and refusing to call a spade a spade?

In the spirit of modern philosophy, I seem to find that the problem of evil turns on a certain unfortunate way of using the term 'evil'. I hope to show clearly in what way this is so and how, on more careful analysis, one can preserve both the goodness and omnipotence of God without sacrificing the vocabulary necessary to an adequate description of the various horrors which history has furnished (and continues to furnish) us.

Before proceeding, let us note that this is not an article on the existence of God. The problem I pose is essentially a logical one – the question of reconciling the seemingly contradictory character of attributing both goodness and omnipotence to any God that exists. I will not bother to punctuate my article with conditional phrases of 'if God exists, then . . . ,' and the reader is invited to insert them or not according to his personal convictions. The point is that I am begging no question in refusing to discuss here the existence of God.

Analysis

Let us now return to the argument which constitutes the problem of evil, stating all of its premises explicitly so that a precise, logical analysis may be obtained:[1]

$$(Ey)[Ev\,(y)] \tag{1}$$
'There is at least one thing which is evil.'

$$(x)[Ev\,(x) \supset -Gd\,(x)] \tag{2}$$
'No matter what thing we choose, if it is evil,
then it is not good'; more briefly said: 'Nothing
which is evil is good.'

Notice that statement (2) is minimal in the assumptions it makes about the relationship between good and evil, because it does not identify goodness with nonevil. By the laws of logic, we can of course infer from (2) that if something is good then it is not evil, and this we certainly want to be true. But we cannot infer that if something is not evil then it is good. Hence, goodness can be thought of as a positive quality, something more than the mere absence of evil. The logical point here is that we do not have to decide whether to identify goodness with nonevil for the purposes of this discussion. If we obtain a contradiction involving the assumption (2), then we will *a fortiori* be able to obtain a contradiction from the stronger assumption:

$$(x)[Ev\,(x) \equiv -Gd(x)] \tag{2$'$}$$
'Anything is evil if and only if it is not good.'

We continue:

$$(x)(y)\{[Rsp\,(x,y) \wedge Ev\,(y)] \supset Ev\,(x)\} \tag{3}$$
'If one is responsible for something which is
evil, then one is evil'; more simply: 'To be
responsible for evil is to be evil.'

Note that 'responsible' is a relative predicate 'x is responsible for y' and not an absolute predicate such as 'evil'. The extension (set of satisfying values) of a relative predicate is a class of ordered pairs of objects, while the extension of an absolute predicate is a class of objects.

$$(x)[Pw(x) \supset (y)Rsp(x,y)] \qquad (4)$$
'If something is all-powerful, then it is
responsible for everything that exists.'

$$(x)[Cr(x) \supset Pw(x)] \qquad (5)$$
'No matter what thing we choose, if it is God
(symbolized as Cr for "creator"), then it is
all-powerful.'

From premises (1)–(5), all assumptions on which the 'problem of evil' is based, we can conclude, using only the laws of (modern) logic, that

$$(x)[Cr(x) \supset -Gd(x)] \qquad (6)$$
'No matter what thing we choose, if that thing
is creator of the universe, then it is not good.'

The formal deduction is exhibited below. The reader can skip the details of the formal deduction and accept the conclusion or give for himself an informal deduction if he chooses.

In the following deduction, the bracketed 1 indicates dependence on the hypothesis of line 1 for the lines of the deduction where the bracketed 1 is displayed. The notations H, $e\forall$, MP, eE, eH, and $i\forall$ stand for 'hypothesis', 'eliminate universal quantifier', '*modus ponens*', 'eliminate existential quantifier', 'eliminate hypothesis', and 'introduce universal quantifier', respectively.

[1]	1. $Cr(x)$	H
	2. $Cr(x) \supset Pw(x)$	$e\forall$, premise (5)
[1]	3. $Pw(x)$	1,2, MP
	4. $Pw(x) \supset (y)Rsp(x,y)$	$e\forall$, premise (4)
[1]	5. $(y)Rsp(x,y)$	3,4, MP
	6. $(Ey)Ev(y)$	premise (1)
	7. $Ev(a)$	6, eE (a, some new constant)
[1]	8. $Rsp(x,a)$	5,$e\forall$
	9. $[Rsp(x,a) \wedge Ev(a)] \supset Ev(x)$	$e\forall$, premise (3)
[1]	10. $Ev(x)$	7,8,9, tautology, MP
	11. $Ev(x) \supset -Gd(x)$	$e\forall$, premise (2)

[1]   12. $-Gd(x)$ 10,11, MP
           13. $Cr(x) \supset -Gd(x)$ 1,12, eH
           14. $(x)[Cr(x) \supset -Gd(x)]$ 13, $i\forall$.

Each of our premises has been used in obtaining the conclusion.

If we wish to add the explicit premise that God exists, then we will have

$$(E!x)\,Cr(x) \qquad (7)$$
'There exists one and only one God.'

We can then state, using the description operator,

$$-Gd\,[\iota xCr(x)] \qquad (8)$$
'God is not good.'[2]

Statement (8) is provable if (7) is added as a premise.

Whether or not we make the explicit hypothesis (7), the logical point is the same: The assumption of the existence of a God leads to the conclusion that he is not good.

If we take as premises (1)–(4), replacing (5) by

$$(x)\,[Cr(x) \supset Gd(x)] \qquad (9)$$
'Whatever we choose, if it is God, then it is good.'

we can formally deduce the conclusion

$$(x)[Cr(x) \supset -Pw(x)]$$
'Whatever thing we choose, if it is God, then
it is not all-powerful.'

We do not furnish the details of the deduction, letting the above serve as an example.

If, now, we suppose (1)–(4) and replace (5) by

$$(x)\{Cr(x) \supset [Gd(x) \wedge Pw(x)]\} \qquad (5')$$
'Whatever thing we choose, if it be God, then
it is good and all-powerful,'

then we can formally deduce the conclusion

$$(x)\{Cr(x) \supset [-Pw(x) \wedge Pw(x)]\} \qquad (10)$$
'Whatever thing we choose, if it is God, then it is
both all-powerful and not all-powerful.'

From this we have immediately

$$-(Ex)\,Cr\,(x)\qquad\qquad(11)$$
'There is no God.'

Thus, if we add to the set (1)–(4) and (5') the further premise (7) that there is a God, we immediately obtain a contradiction. Thus, God, if he exists, cannot be both good and all-powerful on pain of formal contradiction. Notice again that we have never used the stronger assumption (2').

Explicitly, the set of premises which leads to formal contradiction is the set (1)–(4), (5'), and (7). Let us examine these one by one to determine likely candidates for rejection in order to avoid contradiction.

As we have already stated, we are not interested in the rejection of (7) in this article. Of course, the fact that the above set of statements is contradictory has sometimes been used precisely as an argument for the rejection of the existence of God. But any reasonable solution to the problem which avoids the rejection of (7) will show that such an argument is inconclusive.

The refusal to reject (5') has already been seen as the heart of the problem we are attacking. Our precise intention here is that we shall not take this way out.

Rejection of (2) seems weak, since this would appear to be the least prejudicial way of asserting the relationship between good and evil, as we have already noted.

Rejection of (4) is also unsatisfactory, since this is almost a definition of terms. To be all-powerful means precisely to control everything, thus to be responsible for everything. Man is not all-powerful precisely because there exist things (the universe, for example) for which he is not responsible.

One could argue for a rejection of (3), which says that to be responsible for evil is evil. There are those who have argued in the vein that this is not necessarily so. It has been said, for example, that God 'uses evil' for good purposes. Some have even waxed eloquent, pointing out that the very proof of Godliness is that God is so powerful, clever, or what have you that he can use evil for good.

There does indeed seem to be a grain of truth in this type of argument. We often observe processes in life in which something we call evil works toward an end which we judge desirable and good. It can be pointed out that suffering often entails growth and development, serving as a stimulus to organisms to seek higher and more creative forms of adaptation.

What weighs most heavily against this argument is the equivocation of terms it seems to involve. Can that which is evil really lead to good? If something leads to good, then on what basis do we call it evil in the first place? After all, we may simply be mistaken in calling a particular instance of suffering an evil. Our later realization that the experience resulted in good should occasion the reflection that we were wrong to predicate evil of the suffering to begin with, not that something which was intrinsically evil has magically changed to good!

In short, an evil, whatever else it may be, must be something that, by its very nature, does not tend toward good ends. The fact is that most life situations involve a mixture of factors, some of which we judge good and others evil. If we are consistent in our use of these terms, we must suppose that the good which results from a given situation results from the good involved and that the result would have been even better had the evil involved not been there at all. That a God could produce some good results where evil is involved does not imply that it was the evil which contributed to the good result. The good which results from a situation must result in spite of the evil involved and not because of it. Otherwise, our use of the terms 'good' and 'evil' is going to be equivocal.

To sum up, then, evil must by its very nature be something which does not lend itself to good use, and thus to be responsible for evil is to contribute willingly toward the frustration of a certain amount of good. To be responsible for evil is to contribute willingly to a lesser good. It is to be a willing accomplice to the undoing of (a certain amount of) good. And certainly a being who is a willing accomplice to the undoing of good is evil. Thus, rejection of (3) only shifts the philosophical argument to another level and accomplishes nothing.

The above argument for the rejection of (3), as cited above, does seem to have a certain force as an argument for the rejection of (1). We can argue that everything which we call 'evil' tends, from some ultimate and olympian point of view which we do not possess, to work toward good, and thus that evil, in the precise sense we have discussed, that is, in the sense of tending toward the frustration of good, does not exist.

On the other hand, if good exists, then let us identify something which is good and we will certainly discover that some person (perhaps out of ignorance or selfishness) has deliberately attempted to frustrate it. Such acts exist and, since they tend to frustrate good, are evil (and they will hurt at least the authors of such acts). Hence it seems that, if good exists and human freedom is not illusory, then evil must also exist.

Thus, the above argument applied as an argument for the rejection of (1) seems to deny the possibility of good and evil altogether and leaves us with amorality. Again, we have difficulty squaring our philosophical amorality with our value-charged experience of life.

Solution

The solution to the problem lies, I feel convinced, in the observation that the term 'evil', like the term 'responsible', is a relative term. An absolute term (such as 'all-powerful') has a class of objects as its extension (the class of all all-powerful things). It thus divides the ontological universe into two separate parts, those objects which satisfy the term and those which do not (those things which are all-powerful and those which are not). This follows from the logical truth

$$(x)[F(x) \lor -F(x)],$$

where F is any one variable predicate. However, a relative term (such as 'responsible') has a class of ordered pairs of objects as its extension (the class of all pairs $\langle x,y \rangle$ such that x is responsible for y) and does not so divide the universe.

Of course, where F is any relation,

$$(x)(y)[F(x,y) \lor -F(x,y)]$$

is also a logical truth, but this says merely that, no matter what two objects we choose, either they stand in the relation F or they do not.

What we are about, then, is the following: We propose to replace the absolute term

$$Ev(x)$$
'x is evil'

with the relative term

$$Ev(x,y)$$
'x is more evil than y'.

Let us work, rather, with the converse relation

$$Val(x,y)$$
'x is better than y',

understanding that x is better than y if and only if y is more evil than x. We now replace the contradictory set of statements (1)–(4), (5'), and (7) with the following noncontradictory set:

$$(Ex)(Ey) [Val(x,y)] \tag{i}$$
'There exist x and y such that x is better than y
(or, equivalently, y is more evil than x).'

$$(x)(y)[Val\,(x,y) \supset -Val\,(y,x)] \qquad\qquad\text{(ii)}$$

'For any two things x and y, if x is better than y, then y is not better than x.'

$$(x)[-Val\,(x,x)] \qquad\qquad\text{(iii)}$$

'Nothing is better than itself.'

$$(E!x)[Cr\,(x)] \qquad\qquad\text{(iv)}$$

'God exists.'

$$Pw[\iota x Cr(x)] \qquad\qquad\text{(v)}$$

'God is all-powerful.'

$$(y)\{[y \neq \iota x Cr(x)] \supset [Val(\iota x Cr(x),y)]\} \qquad\qquad\text{(vi)}$$

'God is better than every other thing'; in other words, God is the supremely valued thing, the highest good.

$$(x)\{Pw(x) \supset (y)[Rsp(x,y)]\} \qquad\qquad\text{(vii)}$$

The same as (4).

The set of statements (i)–(vii) is clearly consistent. To see this, take as a model the negative integers where Val is the relation 'greater than', the unique object satisfying the predicate Cr is -1, Pw and Cr are both equal to the set whose only element is -1, and Rsp is the relation 'greater than or equal to'. (In fact, the statements clearly have a model in a two-element domain.)

In this set of statements, both the goodness

(in[vi])

and the omnipotence

(in[v])

of God are affirmed. Notice that we no longer have any analogue of (3) in the new set of statements. Let us examine this in more detail.

Premise (3) affirms that to be responsible for evil is evil. This is when we regard 'evil' as an absolute term. We could still obtain a contradiction from the set (i)–(vii) by adding the following statement:

$$(x)(y)\{[Rsp(x,y)] \wedge (Ez)[Val(z,y)] \supset (Ew)[Val(w,x)]\} \qquad (3')$$

'If someone x is responsible for y and there is something z
which is better than y, then there is something w
which is better than x.'

That contradiction follows from (i)–(vii) plus $(3')$ can be seen roughly in this way: By (v) and (vii), God is responsible for everything. By (i), there is a y which is more evil than some x. Since God is responsible for everything, he is responsible for this y. Thus, $(3')$ would require that there be something better than God. But (vi) contradicts this by asserting that God is the supreme good (i.e. is better than every other thing). Roughly, then, we would have a new 'problem of evil' which would go somewhat as follows: God cannot be the supreme good since he is responsible for the fact that there is at least one thing which is more evil than another.

But here the argument for the acceptance of $(3')$, thus forcing the new 'problem', is quite weak. For God is responsible not only for the y that is more evil than x but also for the x which is *better* than y! In short, God is responsible for the fact that some things are better than others. It does not follow in any easily arguable way that God should be held less than supremely good because of this state of affairs.

If we accept a still further hypothesis that humans have a limited but real freedom to choose, then it follows, together with the above, that moral choice is possible. Since some things are better than others, the consequences of moral choices are real. Moreover, by (v) and (vii), God is responsible for this situation.

Suffering (or increased suffering) is often the consequence of wrong moral choice, and one therefore could argue that God is not supremely good because it would have been better for God not to have created this situation. God, since he is all-powerful, could have arranged things otherwise. Let us note, however, that the main logically possible alternatives seem to involve either suppressing the relation Val (amorality again), or suppressing man's freedom, or not creating man in the first place. In fact, all of these logical possibilities amount more or less to the same thing, since it is only the relation *Val* which gives our freedom any meaning or purpose. The freedom to choose among a number of morally indifferent alternatives would be the same as having no freedom, since the result of the 'choice' would not be of any consequence.

On the other hand, the idea that some things are better than others — that some choices lead to relatively good results whereas others lead to relatively bad results — is the very basis of our notion of progress, of growth (both individual and social), and of happiness.

It is obvious that any question can be argued, so the main point here should not be obscured: It is that the burden of proof has now been shifted to the shoulders of those who would argue that God was 'wrong' to allow man the freedom of moral choice. True, we do not see the ultimate end of many of the sufferings we endure, and this may sometimes lead us to curse the freedom which makes us have to suffer. But the alternative of being a dumb automaton (or of not existing at all) seems much more evil, so any argument that this alternative is necessarily a greater good is inconclusive at best. (Nothing, in fact, excludes that even automatons could suffer.) In short, a person can choose to deny the supreme goodness of God on this basis if he chooses, but he cannot feel secure in having done so on such a clear and logical foundation as if our first analysis had been allowed to stand.

I would like to make two observations in closing. The first is this: It is interesting and important that at least one major religion, the Bahá'í Faith, has taken essentially the present solution to the problem of evil.[3] I say that this is important because philosophies are noted for their lack of influence on the public at large while religions are noted precisely for their general influence. That a major religion has avoided the confusion on this issue and assumed a logical stand is thus a good omen.

The second observation concerns the nature of our solution. Notice that, in one sense, our solution harks back to the one first considered in the fourth paragraph of this article, that is, denying the existence of evil. Of course, we have not rejected evil but rather 'evil'. We have not rejected the existence of the moral dimension but rather the term 'evil' as an absolute term. My question is this: Could other thinkers, such as Leibniz, who were led to deny the existence of evil really have been attempting to formulate something like the present solution?

Our analysis has rested heavily on the logic of relations, and this was developed only late in the nineteenth and early in the twentieth century by De Morgan, Frege, Schröder, and Russell in Europe and by C.S. Peirce in America. Hence, the present way of escaping the dilemma was denied those who thought about the problem before modern times, simply because the necessary vocabulary was not yet common philosophical currency.

The question is particularly poignant in regard to Leibniz, for it is well known that it was he who first conceived of the possibility of a logical calculus and even made unsuccessful attempts to develop it. Could he have intuitively conceived of an analysis resembling the present one and yet have remained unable to express it adequately due only to the above-mentioned lack of vocabulary (the logic of relations)? For my part, I like to think so, for certainly this is more reasonable than to assume that the thought of this incomparable genius was vulnerable to the amusing but philosophically naïve attack of Voltaire's *Candide*.

5

Science and the Bahá'í Faith

Part of the difficulty involved in attempts to understand and clarify the relationship between religion and science is that the nature of religion seems much less clearly defined than that of science. Is religion primarily a cognitive activity like science, or is it more akin to an aesthetic or emotional experience? If religion is seen as primarily cognitive, then the main problem seems to be that of reconciling the application of scientific method to religion. In particular it is often felt that this is difficult to do without falsifying either the nature of scientific method or else the global, subjective, mystic character of religion. On the other hand, viewing religion as primarily noncognitive appears ultimately to relegate religion to an unacceptably secondary and inferior status in the range of human activities. It becomes very difficult to attribute any objective content to religious belief and to religious moral imperatives. These latter are seen at best to be expressions of various subjective, emotional, essentially irrational (and per-

This essay was originally published in *Bahá'í Studies*, vol. 2 (September 1977), © 1977, 2nd rev. ed. 1980 by the Association for Bahá'í Studies. It was reprinted in a slightly revised form in *Zygon*, vol. 14, no. 3 (September 1979), 229–53.

haps illegitimate and illusory) yearnings and desires on the part of a more or less general segment of mankind.

The Bahá'í Faith, founded in 1844 in Persia under extraordinary circumstances, is significant among the religions of the contemporary world in its clear statement both of the nature of religion itself and of the applicability of scientific method to religion. In a summary description of basic Bahá'í beliefs Shoghi Effendi (1897–1957) affirms:

The Revelation proclaimed by Bahá'u'lláh, His followers believe, is divine in origin, all-embracing in scope, broad in its outlook, *scientific in its method*, humanitarian in its principles and dynamic in the influence it exerts on the hearts and minds of men. The mission of the Founder of their Faith, they conceive it to be to proclaim that religious truth is not absolute but relative, that Divine Revelation is continuous and progressive, that the Founders of all past religions, though different in the non-essential aspects of their teachings 'abide in the same Tabernacle, soar in the same heaven, are seated upon the same throne, utter the same speech and proclaim the same Faith'. His Cause, they have already demonstrated, stands identified with and revolves around, the principle of the organic unity of mankind as representing the consummation of the whole process of human evolution. This final stage in this stupendous evolution, they assert, is not only necessary but inevitable, that it is gradually approaching, and that nothing short of the celestial potency with which a divinely ordained Message can claim to be endowed can succeed in establishing it.

The Bahá'í Faith recognizes the unity of God and of His Prophets, upholds the principle of an unfettered search after truth, condemns all forms of superstition and prejudice, teaches that the fundamental purpose of religion is to promote concord and harmony, *that it must go hand-in-hand with science*, that it constitutes the sole and ultimate basis of a peaceful, an ordered and progressive society.[1]

Further, the essentially cognitive nature of religion is affirmed by the founder, Bahá'u'lláh (1817–1892), in language such as:

First and foremost among these favors, which the Almighty hath conferred upon man, is the gift of understanding. His purpose in conferring such a gift is none other except to enable His creature to know and recognize the one true God – exalted be His glory. This gift giveth man the power to discern the truth in all things, leadeth him to that which is right, and helpeth him to discover the secrets of creation. Next in rank, is the power of vision, the chief instrument whereby his understanding can function. The senses of hearing, of the heart, and the like, are similarly to be reckoned among the gifts with which the human body is endowed . . . These gifts are inherent in man himself. That which is preeminent above all other gifts, is incorruptible in nature and pertaineth to God Himself, is the gift of Divine Revelation. Every bounty conferred by the Creator upon man, be it material or spiritual, is subservient unto this.[2]

In other words, from the Bahá'í viewpoint religion is basically a form of knowing, the object of knowledge (or basic datum) of which is the phenomenon of revelation. The other mystic and emotional aspects of religion also are affirmed in the Bahá'í Faith, but still the Faith is proclaimed to be 'scientific in its method'. Another essential aspect of religion is that of action or 'good works'. Still, 'Abdu'l-Bahá (1844– 1921), son of Bahá'u'lláh and designated interpreter of his father's revelation, affirms the primacy of knowledge with respect to action: 'Although a person of good deeds is acceptable at the Threshold of the Almighty, yet it is first "to know", and then "to do". Although a blind man produceth a most wonderful and exquisite art, yet he is deprived of seeing it. . . . By faith is meant, first, conscious knowledge, and second, the practice of good deeds.'[3] He defines religion as 'the essential connection which proceeds from the realities of things' or 'the necessary connection which emanates from the reality of things', again ascribing objective, cognitive content to it.[4]

The problem with all of this is that to affirm something as true does not necessarily give us an understanding of how or why it is true. My purpose in this essay then is to discuss the religion-science conflict from a Bahá'í viewpoint with the specific goal of explicating the above affirmations. It is my hope that such an effort may prove of interest and profit to those of any religious background or viewpoint.

The Nature of the Religion-Science Conflict

At the heart of the conflict between science and religion is that two essentially different views of man are associated respectively with each, at least in the popular view. In the one instance man is seen as a superevolved animal, a chance product of a material thermodynamic system. In the other he is seen as a spiritual being, created by God with a spiritual purpose given by God. Of course conflicting views of the nature of man are as old as thought itself and certainly predate the period of modern science. However, it is only in the modern period that the materialistic view has become linked to a prestigious and highly efficient natural science. The prestige of science forces people to take seriously any pronouncement that is put forth in its name.

All of this contrasts sharply with the premodern period in which the materialistic view was just one among many competing views and had no particular natural or obvious superiority over others. People simply could discredit or disregard the materialistic viewpoint without feeling any pangs of conscience or without feeling threatened.

In sum then I am suggesting that the conflict between religion and science is due essentially to the two qualitatively different views of man which are associated respectively with them, that the force of the materialistic view associated with modern science is due not to any inherent philosophical superiority of that view but rather to the immense prestige of the science in the name of which the materialistic view is put forth and that this prestige of science is due essentially to its evident technological productivity and efficiency.

One may ask in turn to what the efficiency and productiveness of modern science is due, and I believe that here there is one basic answer: scientific method. It is the method of science which has led to such remarkable results and thus to the present situation. Although some thinkers have tried to attribute the success of scientific method to one aspect or another of Western culture or religion, it is now abundantly clear that modern scientific method can be practiced with success independently of any particular religious or cultural orientation.

Indeed we can say that science as an activity is characterized by its method, for the immense diversity of domains which are now the object of scientific study defies any intrinsic character-ization in terms of unity of content. The unity of science is its method.

The importance of religion on the other hand derives precisely from its goal and its contents rather than its method. Religion treats of questions which are so fundamental for us that every human being is obliged to realize the importance of answering them. Some of these questions concern the purpose of man's existence, the possibility of life after death, the possibility of self-transcendence, the possibility of contacting and living in harmony with a higher spiritual consciousness, the meaning of suffering, and the existence of good and evil.

Once we realize that the basis of science is its method and that the basis of religion is its object of study, the essential move toward resolving the religion-science controversy seems obvious and logical: Apply scientific method within religion. But, as I already have noted, there is widespread feeling that this is not truly possible. Thus each side remains with its view of the nature of man and with a feeling that a reconciliation is not possible.

It seems to me, however, that the conviction of the impossibility of applying scientific method to religion rests on several misconceptions both of the nature of scientific method and of the nature of religion. The ensuing discussion, though clearly incomplete, attempts to identify the sorts of misunder-standing involved.

The Nature of Scientific Method

Science is, first of all, knowledge. Moreover, it is human knowledge because it is humans who do the knowing, and the nature of human knowledge will be determined by the nature of human mental faculties. Of course every human being on earth knows things and uses his mental faculties in order to attain this knowledge. What distinguishes the scientific method of knowing, it seems to me, is the systematic, organized, directed, and conscious nature of the process. However much we may refine and elaborate our description of the application of scientific method in some particular domain such as mathematics, logic, or physics, this description remains essentially an attempt on our part to bring to ourselves a fuller consciousness of exactly how we apply our mental faculties in the course of the epistemological act within the given domain. I offer therefore this heuristic definition of scientific method: Scientific method is the systematic, organized, directed, and conscious use of our various mental faculties in an effort to arrive at a coherent model of whatever phenomenon is being investigated.

In a word, science is self-conscious common sense.[5] Instead of relying on chance happenings or occasional experiences, one systematically invokes certain types of experiences. This is experimentation (the conscious use of experience). Instead of relying on naïve reasoning, one formalizes hypotheses explicitly and formalizes the reasoning leading from hypothesis to conclusion. This is mathematics and logic (the conscious use of reason). Instead of relying on occasional flashes of insight, one systematically meditates on problems. This is reflection (the conscious use of intuition).[6]

The practice of this method is not linked to the study of any particular phenomenon. It can be applied to the study of unseen forces and mysterious phenomena as well as to everyday occurrences. Failure to appreciate the universality of scientific method has led some to feel that science is really only the study of material phenomena. This narrow philosophical outlook, plus the historical fact that physics was the first science to develop a high degree of mathematical objectivity, has led to a common misconception that scientific knowledge is inherently limited only to physical reality.

It should be stressed also that the scientific study even of material and concretely accessible phenomena involves a heavily theoretical and subjective component. Far from just 'reading the facts from the book of nature', the scientist must bring an essential aspect of creative hypothesis and imagination to his work. Science as a whole is underdetermined by experience, and there are often many different possible models to explain a given phenomenon. The scientist must therefore not only find out how things are but also imagine how things might be. Developments in all branches of science during this century have led to an increasing awareness among scientists and philosophers of the vastness of this subjective input into science.

Another feature of scientific knowledge is its relativity. Because science is the self-conscious use of our faculties we become aware that man has no absolute measure of the truth. The conclusions of scientific investigations are always more or less probable. They are never absolute proofs.[7] Of course if a conclusion is highly probable and its negation highly improbable we may feel very confident in the results, especially if we have been very thorough in our investigation. But realization and acceptance of this essential uncertainty and relativity of our knowledge are important, for the exigencies of the human situation are often such that we are forced to act in some instances before we have had time to make such a thorough investigation. It therefore behooves us to remain constantly alert to the possibility that in fact we may be wrong.[8]

Let us note in passing that a similar view of scientific method is expressed in several places in Bahá'í writings. In a talk delivered at the Green Acre Institute in Eliot, Maine, in 1912 'Abdu'l-Bahá discusses the methods of knowledge or criteria of judgment available to man: 'Proofs are of four kinds; first, through sense-perception; second, through the reasoning faculty; third, from traditional or scriptural authority; fourth, through the medium of inspiration. That is to say, there are four criteria or standards of judgment by which the human mind reaches its conclusions.[9] 'Abdu'l-Bahá then discusses each of these criteria and shows why it is fallible and relative.[10] He then continues:

Consequently it has become evident that the four criteria or standards of judgment by which the human mind reaches its conclusions are faulty and inaccurate. All of them are liable to mistake and error in conclusions. But a statement presented to the mind accompanied by proofs which the senses can perceive to be correct, which the faculty of reason can accept, which is in accord with traditional authority and sanctioned by the promptings of the heart, can be adjudged and relied upon as perfectly correct, for it has been proved and tested by all the standards of judgment and found to be complete. When we apply but one test there are possibilities of mistake.[11]

In still another passage 'Abdu'l-Bahá explains the relativity of man's knowledge:

Knowledge is of two kinds: one is subjective, and the other objective knowledge; that is to say, an intuitive knowledge and a knowledge derived from perception.

The knowledge of things which men universally have, is gained by reflection or by evidence: that is to say, either by the power of the mind the conception of an object is formed, or from beholding an object the form is produced in the mirror of the heart. The circle of this knowledge is very limited, because it depends upon effort and attainment.

But the second sort of knowledge, which is the knowledge of being, is intuitive, it is like the cognisance and consciousness that man has of himself.

For example, the mind and the spirit of man are cognisant of the conditions and states of the members and component parts of the body, and are aware of all the physical sensations . . . This is the knowledge of being which man realises and perceives; for the spirit surrounds the body, and is aware of its sensations and powers. This knowledge is not the outcome of effort and study; it is an existing thing, it is an absolute gift.[12]

'Abdu'l-Bahá then explains that the Manifestations, or revelators, are distinguished from ordinary men in that they have the subjective (intuitive) knowledge of all things: 'Since the

Sanctified Realities, the universal Manifestations of God, surround the essence and qualities of the creatures, transcend and contain existing realities and understand all things, therefore their knowledge is divine knowledge, and not acquired: that is to say, it is a holy bounty, it is a divine revelation.'[13] It is this unique consciousness of the Manifestations which according to him enables them to be the focal point of man's knowledge of God.

In yet another passage 'Abdu'l-Bahá puts the matter thus: 'Know that there are two kinds of knowledge: the knowledge of the essence of a thing, and the knowledge of its qualities. The essence of a thing is known through its qualities, otherwise it is unknown and hidden. As our knowledge of things, even of created and limited things, is knowledge of their qualities and not of their essence, how is it possible to comprehend in its essence the Divine Reality, which is unlimited? . . . Knowing God, therefore, means the comprehension and the knowledge of His attributes, and not of His Reality. This knowledge of the attributes is also proportioned to the capacity and power of man; it is not absolute.'[14]

I will try to sum up, however inadequately, the epistemological implications of these passages in this way: Human knowledge is the truth which is accessible to man, and this truth is relative because man the knower is relative, finite, and limited. There is an absolute reality underlying the multifaceted qualities and experiences accessible to man, but direct access to this reality or direct perception of it is forever beyond man's capabilities. His knowledge is therefore relative and limited only to the knowledge of the various effects produced by this absolute reality (the Manifestations being one of the most important of these effects). However, if man uses systematically all of the various modes of knowledge available to him, he is assured that his knowledge and understanding, such as they are on their level, will increase.[15]

Positivism and Existentialism

The main purpose of this brief discussion of scientific method is to suggest that a misconception of the nature of scientific method — namely, that it is applicable only to more or less concretely accessible material phenomena and only in a relatively narrow way — has led to the general conclusion on the part of many religionists and scientists that scientific method is not applicable to religion.[16] Depending on what further assumptions are made, one is led to two basic positions which I have labeled positivism and existentialism. There are many variants to each position, and so these labels must be understood in a very general, heuristic way.

On the one hand we may add to the narrow view of scientific method the assumption that scientific method (so construed) is the only valid method of knowledge. One then concludes that religion is not a form of knowledge at all but rather an institutionalized form of superstition, emotionalism, fanaticism, togetherness, or what have you. On the other hand we may conclude that there are methods of knowledge other than the scientific one which are appropriate to religion. Religion in this view is so deeply private, mystical, and subjective as to be 'beyond' scientific method. It is of course the first of these views that I have labeled 'positivism' and the second 'existentialism'. I would like now to discuss briefly each of these positions in an attempt to show exactly why I hold them to be mistaken.

Basically the positivistic position regards religion as too hopelessly lacking in objectivity to be accessible to scientific treatment. It is true of course that the subject matter of religion is more complex than that of, say, physics because it includes more parameters. In the same way biology is more complex than physics, psychology more complex than either, and religion the most complex of all. In this sense religion is indeed more 'subjective', for the presence of many more parameters makes objectivity harder to obtain since the effort to make all parameters explicit is correspondingly much greater. Indeed this is quite clearly reflected in the historical development of science in which first physics was developed to a fairly high level of objectivity, followed by chemistry, then biology, and now increasingly psychology and sociology.

But it is important to realize, as I mentioned in the foregoing, that there is an essential part of subjectivity involved in the application of scientific method in any context. Suppose, for example, that we try to eliminate the subjective element of the notion 'red' by agreeing that the term shall be applied only to those objects which give a reading of thus and so on a spectroscope. Once this agreement is made we may still argue sometimes about whether or not the needle really is quite on thus and so, and the unbeliever will go away saying that the definition was all wrong in the first place.

Thus subjectivity is involved in science even on the most basic, observational level. It is obviously involved even more on the theoretical level where the entities discussed are not directly observable and where many of the statements are not directly testable empirically. Though parts of the total context of science may involve highly articulated objectifications, the ultimate roots of understanding lie always in collective human subjectivity, and so there is always 'room for argument'.

Besides appealing to explicit conventions, formal logic, and the like, positivists have tried to discredit the application of scientific method in religion by insisting on public verifiability as an essential aspect of scientific method. However, a little reflection will show easily that this restriction is arbitrary and in no wise a criterion of scientific method. I offer the following paradigm as an illustration of this point.

A biologist looks through a microscope in his laboratory, sees a certain configuration, and exlaims: 'Aha, at last I have the evidence that my theory is correct!' Question: How many people in the world are capable of looking at the configuration and verifying the findings of the biologist? Answer: Very few, almost none, probably only a few specialists in his field. The fact is that the biologist will publish his findings, and a few other qualified individuals will test his results, and if they seem confirmed the scientific world at large will accept the theory as verified. Although the positivist might concede this, he would say: 'But if an individual did go through the years of training necessary to understand everything the biologist knows, then the individual could verify the statement. Thus, I admit the statement is not practically verifiable by the public, but it is theoretically verifiable.' But even this is not enough. The fact is that the positivist will be constrained to admit that a great many people may be unable, through lack of intelligence or mental proclivity, ever in theory to validate the result. The fact is that the findings are not verifiable by the public at all. The findings can be verified only by individuals capable of assuming and willing to assume the point of view of the researcher. In most instances this group is a very select one indeed, drawn from those who are members of a community of understanding and who participate in a certain framework of interpretation applied to all those subjective experiences which fall within a certain category. More will be said of this later.

At bottom the criterion for truth in science is pragmatic. 'Does it work the way it says it will?' is the question to be answered. If the theory says that such and such a thing must happen, then does it happen? It is by repeated application of this pragmatic criterion, interlaced with intervening theory, that we gradually build up a model of reality, a collection of true statements. We may formulate a general criterion of scientific truth as follows: We have a right to accept a statement as true when we have rendered it considerably more acceptable than its negation. Proof in scientific terms means nothing more than the total process by which we render a statement acceptable by this criterion. Such a proof remains always relative, for it depends on the total context of the statements involved, the implicit and explicit conventions concerning the meaning and operational use of symbols, the experiential component of these statements, and so on. All of these things have their ultimate roots in human subjectivity and are therefore liable to possible revision in the future.

In practice of course it often happens that revision comes either from strikingly new and different experiences which demand that we revise our conceptual framework in order to account for them or from some unexpected conclusions which are deduced within the framework itself and which contradict known experiences (the most radical case being that of logical contradictions). But nothing excludes the possibility that revision may come from some subtle interaction of all of these factors in a way which is totally inconceivable to us at present.

In short, I maintain that any sort of formulaic, pseudo-objective characterization of scientific method such as that attempted by various positivistic-minded philosophers cannot truly capture scientific method.[17] Our description of scientific method must remain scientific, that is, pragmatic, relative, open, etc.

Without any such closed, exclusive formula characterization of scientific method there is no basis on which to exclude the application of scientific method to religion. Of course this does not mean that everything that passes for religion is scientific; nor does it allow us to say what we will find if we do apply scientific method to religion. My essential contention is simply that no known positivistic formulations of or restrictions on the nature of scientific method which exclude *a priori* the applicability of scientific method to religion seem to be justified by the nature of scientific method itself. Furthermore, the nature of scientific method does not appear to lend itself to such formulations or restrictions.

The existentialist position derives its character more from its view of religion than from its view of scientific method.

The existentialist might well accept, even readily, that scientific method cannot be applied to religion. But such a contention would not bother him (as it does me) because it only serves to heighten the difference and cleavage between science and religion. For him the very importance of religion derives from its being unsystematic, even chaotic, subjective, private, uncommunicable, emotional, etc. For him the knowledge that religion brings is a mystic or occult knowledge, communicable only to a limited extent and primarily through myth, symbol, art, and other forms of nonverbal activity.

One extreme form of this position would be to accept completely the positivistic contention that religion is not a form of knowledge and to view religion primarily as an aesthetic experience of some sort. Otherwise if religion is viewed as a form of knowledge it is a form totally different from science, with its own methodology (or lack of methodology), symbols, and experiences.

Perhaps in the last analysis the difference between the existentialist and the positivist lies not so much in their respective views on the nature of religion and of science as in their difference in attitude toward these perceptions. The positivist values science above religion and sees his narrow interpretation of scientific method, with the consequent exclusion of religion, as purifying science from the unwanted trash of emotionalism and irrationality. The existentialist values religion above science and is just as glad to see religion separated from what he feels to be the soul-stultifying dryness, uniformity, formalism, and mechanization of science. While the positivist is impressed primarily by the efficiency and achievements of science, the existentialist is impressed by the potential richness of subjective experience. This richness he sees as constituting that which is most truly human and which deserves to be most thoroughly and strenuously developed in man. Since, as he supposes, scientific methods cannot be used to develop this richness, religion must develop methods of its own different from those of science. It is to the development of such methods that the existentialist bends his efforts, and it would never occur to him to try to reconcile religion and science, something which he would regard as impossible in any case.

My sketch here of what I have labeled the existentialist position is consciously exaggerated at some points, but the logical thrust is clear: The existentialist grants that science cannot be applied to religion, that religion is peculiarly subjective and mystical in a way that makes it necessarily unsystematic and thus inaccessible to science, and he values this subjective aspect of religion above science and its method. He is therefore not upset by the cleavage between religion and science (except that he may have existential difficulties living in a world which is currently dominated by science and its fruits!).

Now I am as impressed as anyone by the richness of subjective experience, and I certainly feel that if the practice of science, or anything else, is going to lead ultimately to a progressive impoverishment of it, then such practice is de-humanizing and should be abandoned. But I feel that the existentialist position and its variants fall into their particular view of internal experience only by neglecting seriously the collective and social dimension of religion, in short, by considering religion as something which is purely internal to the individual. It is only within such a framework that the subjective aspect can be isolated from the rest of religion and made to seem inherently separate from other types of subjective experience, in particular from that involved in the practice of science itself.

We already have had occasion, in the foregoing, to appreciate the fact that subjective experience is involved intimately and irrevocably in the practice of science at all levels. Clearly it is more reasonable then to view subjective experiences as being ranged on some sort of continuum from less intense to more intense, or from less profound to more profound, or yet some other characterization. As different as may be the experience of seeing a red object on the one hand and that of mystical ecstasy on the other, they are generically instances of subjective experience before they are specifically anything else. Moreover, the practicing scientist and the mystic, when confronted with the problem of building and communicating conceptual models of their experience, face essentially the same logical difficulty on their level of experience. For everyone, including the scientist, knows that no amount of explication, verbal or otherwise, can ever exhaust all of the subjective richness of the experience of 'red'. Our previous example of the spectroscope shows the nature of the problem involved, and we must further remember that during the long years of science's evolution such sophisticated conventional devices were not at hand.

Science has overcome this barrier by creating a community of understanding. Each individual scientist must undergo training of a sort which enables him to participate in the validation of the subjective experience of other members of the scientific community when this experience falls within a certain range determined by the nature of the particular scientific discipline in question. As we have seen in the example of the biologist and his microscope, subjective experience is never publicly verifiable. It is verifiable only by those capable of assuming and willing to assume the point of view of the one who has the experience. By maintaining a growing discipline of education and training in science a community of qualified individuals capable of assuming and willing to assume a certain point of view is evolved. This community generates a framework of interpretation for the individual practicing scientist, and it is this framework of interpretation which alone enables his own work, however brilliant or insightful, to become truly illuminating. No matter how far above the common lot of scientists an Einstein or a Newton may be, he can function significantly only in the context of such a community of understanding. If these same individuals had been born in a desert or in a tropical rain forest, their subjective experience would have fallen within another framework of interpretation and would certainly not have had the same result (though it may have been just as illuminating in its own context).

This model of the objectification of internal experience through creating a community of understanding and a consequent framework of interpretation is borne out by observation and experience not only of the history and development of science but also of individuals. For example, case histories of individuals blind from birth who were given sight after reaching maturity indicate, as one would expect, that perception is not immediate but has to be painfully and slowly learned. Their first experience is a chaos of sensations with no discernible objects, forms, etc. Gradually, through participation in the framework of interpretation given by the community, perception is born, and order is brought out of chaos.[18]

The neglect of the social dimension of religion is only one aspect of the weakness of the existentialist position. Another aspect comes into focus when we further examine the comparison between the scientific view of subjective experience and the existentialist view. While our discussion of scientific method has led us to acknowledge a certain irreducibility of the subjective input into the epistemological act, it is nevertheless equally clear that our experience, however subjective, of anything, say a red object, is still an experience of something. Even the chaos of sensation that the previously sightless person experiences is a reaction of his subjectivity to something 'out there'. It is not simply the mind's experience of itself (which might be likened to the sensations of images one has during sleep or when one's eyes are closed). But the existentialist glorification of the subjective amounts to treating the internal experience of the individual as the datum of religion. Religious experience is thus not viewed as an experience of anything, at least not anything other than the internal self of the individual.

Insofar as religion is scientific it thus would be indistinguishable from psychology, and this again explains the tendency to emphasize the unsystematic, unpredictable, irrational, mythic, and aesthetic aspects of religious experience, for these are the only aspects which from such a standpoint can be viewed as properly and specifically religious.

If such a view of religion and religious experience is to be refuted one must face and answer the basic question, 'Of what is religious experience an experience?' What is religion about? If scientific method can be applied to religion, then what is the datum of religion? How can we ascribe objective content to religion?

The Bahá'í Faith

The answer which the Bahá'í Faith offers to this central question is, or so it seems to me, particularly cogent, clear, and direct. For Bahá'ís the datum of religion is the phenomenon of revelation. Religion is that branch of knowledge which takes this phenomenon as its special object of study. The objective content of religion derives from this external, phenomenal datum. Religious experience in this view is a response to the spirit and teachings of the revelator or Manifestation.

The Bahá'í Faith offers the scientific hypothesis that revelation is a periodic phenomenon for which the period (i.e. the average time interval between two successive occurrences of the phenomenon) is fairly long.[19] The large number of generations intervening between two occurrences of revelation poses obvious problems for the study of this phenomenon. However, we cannot refuse to study something simply because the study is hard or because the data associated with it are in some instances accessible only with difficulty. Other natural sciences, such as astrophysics, also study periodic phenomena whose periods are much greater than a thousand years and for which the accessibility of data is likewise a problem. Simply, allowances have to be made for the fact that, because of the periods involved, careful records must be kept since the observations which a given individual scientist can make in his lifetime are too limited to form in themselves a basis for the furtherance of the science.

Let us take a brief look at the phenomenon of revelation as it presents itself to us in history, which is man's collective experience.

If we consider the great religious systems of which there still exists some contemporary expression or some historical record, we will see that virtually all of them have been founded by a historical figure, a unique personage. Islám was founded by Muḥammad, Buddhism by Buddha, Christianity by Jesus, Judaism (in its definitive form) by Moses, Zoroastrainism by Zoroaster, and so on. These religious systems have all followed quite similar patterns of development. There is a nucleus of followers gathered around the founder during his lifetime. The founder lays down certain teachings which constitute the principles of his religion. Moreover, each of these founders has made the same claim, namely, that the inspiration for his teachings and his influence was due to God and not to human learning or human devices. Each of these founders claimed to be the exponent on earth of an invisible, superhuman reality of unlimited power, the creative force (creator) of the universe. After the death of the founder, an early community is formed, and the teachings of the founder are incorporated into a book (if no book was written by the founder). And finally a great civilization based on the religious system grows up, a civilization which lasts for many centuries.

All of the statements in the preceding paragraph have high empirical content and low theoretical content. These are a few facts of religious history. Of course they are based on records and observations of past generations. We can try to dispute these records if we choose, but we must be scientific in any approach we make. In particular the records of the older religions are of validity equal to any other record of comparable date. If, for example, we refuse to believe that Jesus lived, we must also deny that Socrates lived, for we have evidence of precisely the same validity for the existence of both men. The records of Muḥammad's life are much more valid historically than these and are probably beyond serious dispute. Moreover, if we choose to posit the unreality of the figures whose names are recorded and to whom various teachings and influence are attributed, we must give at the same time an alternative explanation for the tremendous influence which these religious systems, elaborated in the name of these founders, have had. This is more difficult than we may be inclined at first to believe.

The major civilizations of history have been associated with the major prophetic religious systems. Zoroastrianism was the religion of the 'glory of ancient Persia', the Persia that conquered Babylon, Palestine, Egypt, and the Greek city-states. Judaism was the basis of Hebrew culture, which some philosophers such as Karl Jaspers regard as the greatest in history. Moreover, Jewish law has formed the basis of common law and jurisprudence in countries all over the world. Western culture, until the rise of modern science, was dominated by Christianity. The great Muslim culture invented algebra and preserved and developed the Hellenistic heritage. It was probably the greatest civilization the world had seen until the rise of the industrial revolution began to transform Western culture.

We are, however, very much in the same position with respect to past revelations as we are with regard to any phenomenon of long period. We were not there to observe Jesus or Muḥammad in action. The contemporaries of these people were certainly impressed by them, but these observations were made years ago and are liable, we feel, to embellishments. Even though it may be unscientific to try to explain away the influence of these religious figures, there is still a certain desire to do so. We are put off by some obvious interpolations, and we are not sure just what to accept and what to reject.

Bahá'ís believe that man's social evolution is due to the periodic intervention into human affairs of the creative force of the universe by means of the religious founders or Manifestations. What is most significant is that the Bahá'í Faith offers fresh empirical evidence, in the person of its own founder, that such a phenomenon has occurred. Bahá'u'lláh claimed to be one of these Manifestations, and he reaffirmed the validity of the past revelations (though not necessarily the accuracy of all the details recorded in the ancient books). Here is a figure who walked the earth in recent times and whose history is documented by thousands of records and witnesses. Moreover, the teachings of Bahá'u'lláh are preserved in his manuscripts, and so we are faced with a record of recent date and one about which there can be no serious doubt.

The only way we can judge Bahá'u'lláh's fascinating hypothesis that social evolution is due to the influence of the Manifestations is the way we judge any proposition: scientific method. This is the only way we can judge Bahá'u'lláh's claim to be one of these Manifestations. We must see if these assumptions are consistent with our knowledge of life as a whole. We must see if we can render these assertions considerably more acceptable than their negations. In the case of Bahá'u'lláh we have many things which we can test empirically. Bahá'u'lláh made predictions. Did they come true? Bahá'u'lláh claimed divine inspiration. Did he receive formal schooling, and did he exhibit power and knowledge not easily attributable to human sources? He insisted on moral purity. Did he lead a life of moral purity? In his teaching are found statements concerning the nature of the physical world. Has science validated these? He engaged in extensive analysis of the nature of man's organized social life. Does his analysis accord with our own scientific observations of the same phenomena? He also makes assertions concerning human psychology and subjectivity and invites individuals to test these. Do they work? The possibilities are unlimited.

Of course the same criteria can be applied to other Manifestations, but the known facts are so much less authenticated and so restricted in number that much direct testing is not possible. This does not disturb Bahá'ís because they believe that essentially there is only one religion and that each of the successive revelations is a stage in the development of this one religion. The Bahá'í Faith is thus the contemporary form of religion, and we should not be surprised that it is so accessible to the method of contemporary science. Christianity and Islám were probably just as accessible to the scientific methods of their day as is the Bahá'í Faith to modern scientific method.

This relative inaccessibility of data concerning the older religions should not be taken as in any way lessening their importance or value relative to the Bahá'í Faith. The Bahá'í view is that of the absolute unity of religion, not the superiority of one religion over another for whatever reason.[20] Nevertheless, if one is talking about applying scientific method to religion, problems such as that of the authenticity of ancient records must be faced frankly and seen in their true light. They must be neither exaggerated nor swept under the rug as if they did not matter. Indeed the best of modern biblical scholarship, both Christian and Jewish, has been undertaken in this scientific spirit. If it has resulted in some instances in the undermining of certain traditional beliefs, it has more fundamentally served to clarify and enlighten the faith of truly informed students of religion. If the doubtfulness of a few passages of the Bible has been exposed, the validity of the basic text has been vindicated (e.g. the corroborative version of Isaiah in the Dead Sea manuscripts).

Each religious system has been founded on the faith in the reality of the phenomenon of revelation, and those people associated with the phenomenon felt fully justified in their faith. But as the influence of religion declined and the facts of revelation receded into history the sense of conviction of the reality of the phenomenon subsided, and this was only natural as we have seen. It is therefore important to realize that the Bahá'í Faith offers much more than new arguments about the old evidence for the phenomenon of revelation. It offers empirical evidence for the phenomenon, and it is frank to base itself on this evidence and to apply the scientific method in understanding the evidence. So much is this so that I would unhesitatingly say that the residue of subjectivity in the faith of a Bahá'í is no greater than the residue of subjectivity in the faith one has in any well-validated scientific theory. As in the example of the biologist and the microscope, the findings of a Bahá'í can be verified by anyone willing to assume and capable of assuming the point of view of a Bahá'í.[21]

According to Bahá'u'lláh the social purpose of religion is to create an adequate spiritual basis for the progressive unfolding of an ordered social life for mankind. Indeed, as one examines the history of mankind, one can perceive the gradual ordering and reordering of man's collective life on ever higher levels of unity, each new level maintaining the integrity of the previous ones and at the same time calling forth from the individual a correspondingly greater degree of altruism and other-centeredness. The family, the tribe, the city-state, and the nation can be seen as significant steps in this social evolution. The first two of these successive stages can be identified in large measure with the respective revelations of Abraham and Moses, while the latter is due essentially to Muḥammad, the founder of the nation of Islám.[22] Bahá'u'lláh explains that besides the general mission of renewing the spiritual life of men and society each religion has a specific mission which accomplishes a definite step forward in the total evolution of mankind. He views his own revelation as being the most recent in this succession and as having the unification of mankind as a whole for its specific mission.[23]

As one thinks about this progressive unfoldment of human society one comes to see certain aspects of its mechanism. It is clear that unity on one level can eventually become disunity on another; the unity of the family can coexist with disunity between families, for example. When the new level of unity is first attained it represents a positive step, but the very accretion of power and the increased mastery resulting from the reorganization of society on this higher level can ultimately lead to tensions among these higher-order units themselves. This may happen years or centuries or millennia later, but when it does happen the suffering caused by these tensions becomes increasingly unbearable and serves as one of the factors generating the motivation to accomplish the next stage of unity. That is, the individuals participating in the social system in question develop a strong sense of and a need for the higher unity.[24]

This higher unity is effected not by the suppression of the existing units but by their being harmoniously organized into a still higher unit – the unity of the tribe is the unity of families, the unity of a race that of tribes, the unity of a nation that of races. Indeed the attainment of unity on the lower level has been a necessary prerequisite to its establishment on the higher one. In the same way Bahá'u'lláh envisages world unity as being a unity among nations, with a world government, a world tribunal, a single auxiliary universal language, and a world economic system.

Just as a tree must push its roots deeper as it grows higher, so must each external step forward have an internal concomitant. The individual at each stage must become less self-centered. He must give his loyalty to and identify with an ever-widening circle of his fellow humans. Whereas 'brother' first meant physical brother, it gradually came to mean fellow Jew, fellow brother in Christ, fellow countryman, and ultimately must mean fellow world citizen. There is, in short, a gradual increase in the consciousness of the individual, and it is this new consciousness which alone allows the new unity, the new external step forward, to take place on a spiritual basis. This new depth of individual spiritual awareness also serves to increase the quality of unity at all levels. In this way the creation of the new unity is not a superficial juxtaposition of parts or a purely formal restructuring but a renewal of the whole of the society, indeed the only way the society can be so renewed at that given stage in its development. Thus Bahá'u'lláh teaches that the establishment of world unity will lead to the perfecting and deepening of the quality of life at all levels of society.

This model also explains why we cannot wait for the lower levels of society to become perfect before working on the establishment of world unity (such an objection to the Bahá'í goal of establishing world unity is frequently heard). The interdependence of the part and the whole is too great for such a piecemeal approach to succeed. Bahá'u'lláh explains that mankind is like a body whose cells and organs are the individual human beings and the smaller social units. If the whole body is ill every single cell will be affected in some way. At the same time the whole body suffers to some extent from even a few unhealthy cells.

Thus in the teachings of Bahá'u'lláh there are provisions for the organization and restructuring of society on a world level, and there are provisions for the perfecting of social organization on the local and intermediate levels as well as manifold spiritual aids for the individual in his own effort to spiritualize his life and attain to a new, more universal consciousness.

Indeed the individual aspect of religion is just as essential as the global, social aspect. This individual component was the point of departure for my whole discussion, and so I would like to return to it in closing this essay.

In the Bahá'í worldview the essential purpose of religion for the individual is to provide him with the tools necessary to acquire a true and adequate understanding of his own nature.[25] For Bahá'ís the individual, internal aspect of religion is a direct response to the datum of the Manifestation, his spirit and teachings. It is not simply the mind's experience of itself or some form of autosuggestion. This is why scientific method can be applied even in this aspect of religion. In the Bahá'í Faith the individual component of religion takes the form of daily prayer, communion with God, meditation on the words of Bahá'u'lláh, and a constant effort to express one's developing spirituality through service to mankind. Among the many individual attributes which Bahá'u'lláh mentions as characteristic of the spiritually minded individual are humility, obedience to the will of God, justice, love, abstention from backbiting and criticism of others, regarding others with a sin-covering eye, and preferring others to oneself in all things.

Bahá'u'lláh stresses that personal spiritual development, the experience of self-transcendence, and the mystic sense of union with God — all of which have been described and discussed in the world's mystic literature — are the fruits only of conscious and deliberate search and struggle. They are not haphazard experiences which we can casually cajole from the universe. They must be sought consciously and practiced as diligently as any scientific or academic discipline. Scientific method — the conscious, systematic, organized, and direct use of our mental faculties — must be employed if we are to be successful in developing spirituality.

Of course to say that spirituality must be sought consciously and systematically does not imply that it can be reduced to a formula any more than science itself can be so reduced. 'Abdu'l-Bahá has expressed it simply: 'Everything of importance in this world demands the close attention of its seeker. The one in pursuit of anything must undergo difficulties and hardships until the object in view is attained and the great success is obtained. This is the case of things pertaining to the world. How much higher is that which concerns the Supreme Concourse!'[26]

In contemplating the application of scientific method to individual spiritual practice let us again recall that science never leads to total or absolute objectification of internal experience, for such a thing is simply unobtainable. Moreover, the quality of internal experience involved in the pursuit of spirituality clearly will be infinitely richer than that connected with most other types of activity. In this perspective, emphasis on the aesthetic and the mythic is legitimate, important, and useful, for the gap between any descriptive models of such experience and the experience itself will be correspondingly greater than in other areas, though the basic method remains unchanged. [27]

Religion is primarily a form of knowing but the relativity and limitations of our knowledge will be felt even more keenly here than elsewhere. Indeed it is this self-knowledge, the acute consciousness of these very limitations, which constitutes an important part of our knowledge of God. One of the profoundest truths that the mystic discovers is that the ultimate goal is not to comprehend but to be comprehended. The deepest knowledge is attained by the profoundest awareness of our own relative ignorance. Bahá'u'lláh expresses this important truth:

Consider the rational faculty with which God hath endowed the essence of man. Examine thine own self, and behold how thy motion and stillness, thy sight and hearing, thy sense of smell and power of speech, and whatever else is related to, or transcendeth, thy physical senses or spiritual perceptions, all proceed from, and owe their existence to, this same faculty. . . . Wert thou to ponder in thine heart, from now until the end that hath no end, and with all the concentrated intelligence and understanding which the greatest minds have attained in the past or will attain in the future, this divinely ordained and subtle Reality, this sign of the revelation of the All-Abiding, All-Glorious God, thou wilt fail to comprehend its mystery or to appraise its virtue. Having recognized thy powerlessness to attain to an adequate understanding of that Reality which abideth within thee, thou wilt readily admit the futility of such efforts as may be attempted by thee, or by any of the created things, to fathom the mystery of the Living God, the Day Star of unfading glory, the Ancient of everlasting days. This confession of helplessness which mature contemplation must eventually impel every mind to make is in itself the acme of human understanding, and marketh the culmination of man's development. [28]

Since in the Bahá'í view internal religious experience is not simply the self's experience of itself but is a direct response to the datum of the Manifestation, there is consequently a need for a constantly accessible focal point toward which the individual can turn in his pursuit of these individual spiritual goals. This indeed is one of the reasons for the periodic nature of the phenomenon of revelation. Although something of God's nature can be said to be revealed in every aspect of creation, clearly the force and importance of such a revelation are conditioned by two things, namely, the inherent limitations of the instrument used as a vehicle of revelation and the accessibility to us of the occurrence of revelation.

Man himself is the most highly ordered and subtle phenomenon in all the universe known to man. It thus seems logical that man would be the most nearly perfect (i.e. least limited) instrument available as a vehicle for God's self-revelation, hence the person of the Manifestation.[29] The necessity for the repetition of revelation derives from the condition of accessibility. The length of the period between occurrences, on the other hand, derives from the social nature of religion as described in the foregoing. Simply, it takes a certain time for a Manifestation to become known, his system to become established, and for the specific purpose of his revelation to be accomplished.[30]

Conclusions

I feel that the Bahá'í view of religion is exciting in its fundamental assertion of the objectivity, universality, and accessibility of religion and religious experience to the inquiring mind. The existentialist view of religion, as well as other subjective views, sees religious experience rather as something which cannot (and perhaps should not) be cultivated, practiced, and sought systematically. It must strike like lightning for reasons which are never wholly clear or else as the result of some magical or occult practice. Clearly no experience of such an erratic and unstable nature can ever serve as the basis for a progressive society.

Positivism and its variants limit unduly the application of scientific method and fail to see that the essence of the method can be applied to all phenomena and to all aspects of life, including the spiritual.

The ultimate resolution of the religion-science opposition is based thus on a balance and complementarity between the two, involving a better understanding of the nature and universality of scientific method on the one hand and of the nature and content of that datum which is the phenomenon of revelation on the other. 'Abdu'l-Bahá has expressed admirably the nature of this balance:

Religion and science are the two wings upon which man's intelligence can soar into the heights, with which the human soul can progress. It is not possible to fly with one wing alone! Should a man try to fly with the wing of religion alone he would quickly fall into the quagmire of superstition, whilst on the other hand, with the wing of science alone he would also make no progress, but fall into the despairing slough of materialism . . . When religion, shorn of its superstitions, traditions, and unintelligent dogmas, shows its conformity with science, then will there be a great unifying, cleansing force in the world which will sweep before it all wars, disagreements, discords and struggles – and then will mankind be united in the power of the Love of God. [31]

Works Cited

'Abdu'l-Bahá. *Paris Talks: Addresses Given by 'Abdu'l-Bahá in Paris in 1911–1912.* London: Bahá'í Publishing Trust. 11th ed. 1972.

—— *The Promulgation of Universal Peace.* Wilmette: Bahá'í Publishing Trust, 2nd ed. 1982.

—— *The Secret of Divine Civilization.* Trans. Marzieh Gail. Wilmette: Bahá'í Publishing Trust, 1980.

—— *Selections from the Writings of 'Abdu'l-Bahá.* Haifa: Bahá'í World Centre, 1978.

—— *Some Answered Questions.* Trans. Laura Clifford Barney. Wilmette: Bahá'í Publishing Trust, 1981.

Aristotle. *Works.* Translated under the editorship of W. D. Ross. Oxford: Clarendon Press. Vol. II: *Physica*, 1930; Vol. VIII: *Metaphysica*, 1954.

Avicenna. *Kitab al-Isharat wa-l-Tanbihat.* Ed. J. Forget. Leiden: 1892. Trans. A. Goichon. *Livre des directives et remarques.* Beirut, Paris: 1951.

—— *La métaphysique du Shifa.* Trans. M. Anawati. Montreal: University of Montreal, Institut d'études médiévales, 1952.

—— *Najat.* (Salvation) Cairo: 1938.

Ayer, A. J. *Language, Truth and Logic.* New York: Dover Publications, 1952.

Bahá'u'lláh. *Gleanings from the Writings of Bahá'u'lláh.* Trans. Shoghi Effendi. Wilmette: Bahá'í Publishing Trust, 1971.

—— *Kitáb-i-Íqán: The Book of Certitude.* Trans. Shoghi Effendi. Wilmette: Bahá'í Publishing Trust, 1981.

—— *Tablets of Bahá'u'lláh Revealed after the Kitáb-i-Aqdas.* Haifa: Bahá'í World Centre, 1978.

Bahá'u'lláh and 'Abdu'l-Bahá. *Bahá'í World Faith.* Wilmette: Bahá'í Publishing Trust, 1976.

—— *The Divine Art of Living*. Wilmette: Bahá'í Publishing Trust. Rev. ed. 1974.

Bergson, Henri. *The Two Sources of Morality and Religion*. New York: Scribner's Sons, 1957.

Capra, Fritjof. *The Turning Point*. New York: Bantam Books, 1983.

—— *The Tao of Physics*. Suffolk: Fontana, 1979.

Davidson, H.A. 'Avicenna's Proof of the Existence of God as a Necessarily Existent Being'. *Islamic Philosophical Theology*. Ed. P. Morewedge. Albany, New York: SUNY Press, 1979.

Davies, Paul. *God and the New Physics*. New York: Simon and Schuster, 1983.

Hardy, G.H. *A Mathematician's Apology*. London: Cambridge University Press, 1940.

Hatcher, William. *The Science of Religion: Bahá'í Studies*. Vol. 2, Ottawa: Association for Bahá'í Studies, 2nd ed. 1980.

—— 'Science and the Bahá'í Faith'. *Zygon* 14.3 (1979): 229–53.

—— *The Concept of Spirituality: Bahá'í Studies*. Vol. 11, Ottawa: Association for Bahá'í Studies, 1982.

—— *The Logical Foundations of Mathematics*. Oxford: Pergamon Press, 1982.

Leakey, R. and R. Lewin. *People of the Lake*. New York. Doubleday, 1978.

Merrill, G.H. 'The Model-Theoretic Argument against Realism'. *Philosophy of Science* 47 (1980): 69–81.

Mohr, Hans. *Structure and Significance of Science*. New York: Springer-Verlag, 1977.

Peters, Karl. 'Religion and an Evolutionary Theory of Knowledge'. *Zygon* 17.4 (1982): 385–415.

Plato. *The Republic*. Translated by H.D. Lee. London: Penguin Classics, 1955.

Prigogine, Ilya, and Isabelle Stengers. *Order Out of Chaos*. New York: Bantam Books, 1984.

Putnam, Hilary. 'Realism and Reason'. *Proceedings and Addresses of the American Philosophical Association* 50.6 (1977): 483–98.

Quine, W.V. *Word and Object*. Cambridge, Mass.: M.I.J. Press, 1960.

Rucker, Rudy. *Infinity and the Mind*. Boston: Birkhaser, 1982.

Russell, B. *Why I am Not a Christian*. London: Allen & Unwin, 1958.

Shoghi Effendi. *The World Order of Bahá'u'lláh*. Wilmette: Bahá'í Publishing Trust, 1980.

—— *Directives from the Guardian*. Comp. Gertrude Garrida. New Delhi: Bahá'í Publishing Trust, 1973.

Toynbee, Arnold. *Christianity among the Religions of the World*. New York: Scribner's Sons, 1957.

Universal House of Justice, The. *The Promise of World Peace*. Wilmette: Bahá'í Publishing Trust, 1985.

Weizsäcker, C.F. von. 'Platonic Natural Science in the Course of History'. *Main Currents in Modern Thought*, Vol. 29, 1972.

Wintner, A.F. *The Analytical Foundations of Celestial Mechanics*. Princeton University Press, 1947.

Notes

PLATONISM AND PRAGMATISM

1. I acknowledge the existence of such anti-science viewpoints at this early stage of the essay because I regard them as basically mistaken and do not intend to deal with them again.

2. Speaking of the highest segment of the 'divided line', Plato says, among other things: 'Then, when I speak of the other section of the intelligible part of the line you will understand that I mean that which reason apprehends directly by the power of pure thought . . . The whole procedure involves nothing in the sensible world, but deals throughout with Forms and finishes with Forms.' (*The Republic*, Book VI, No. 511, p. 277.)

3. In a somewhat narrower context, Prof. John Corcoran makes a similar point in considerable detail in his paper 'Platonism and Logicism', Department of Philosophy, S.U.N.Y. at Buffalo, 1977.

4. For a detailed presentation of this notion of scientific method, see my monograph 'The Science of Religion', *Bahá'í Studies*, Vol. 2. See in particular the first and third essays of the monograph where, among other things, this basically pragmatic view of scientific method is found to be quite closely related to views articulated by Bahá'u'lláh, founder of the Bahá'í Faith, and by his eldest son and designated interpreter, 'Abdu'l-Bahá. Indeed, many readers of this monograph have been surprised to find a religion deeply rooted in the Judaic, Christian, and Islamic traditions that articulates and espouses an epistemology so in harmony with modern scientific practice.

In this connection it is interesting to note, although it is in no wise essential to the present argument, that both Bahá'u'lláh and 'Abdu'l-Bahá also appear to espouse some form of Platonic ontology. For example, in his 'Tablet of Wisdom', which deals with a number of ontological questions,

Bahá'u'lláh pointedly praises both Socrates and Plato, stating that it was Socrates, 'the most distinguished of all philosophers . . . who perceived a unique, a tempered, and a pervasive nature in things, bearing the closest likeness to the human spirit, and he discovered this nature to be distinct from the substance of things in their refined form.' (See Bahá'u'lláh, *Tablets of Bahá'u'lláh*, p. 146). In a similar vein, 'Abdu'l-Bahá has affirmed that scientific discoveries and technological inventions pre-exist in the 'invisible world' where they are eventually perceived by the human mind and '[drawn] forth from the unseen into the visible world'. (See 'Abdu'l-Bahá, *Selections from the Writings of 'Abdu'l-Bahá*, p. 170.)

5. Some might feel that this pragmatic conclusion already constitutes sufficient reason to discard indefinitely any recourse to Platonic ontology in connection with scientific method and practice. However, the logical independence of pragmatic method and Platonic metaphysics, which we have mentioned above, shows that such a position cannot be justified on purely logical or pragmatic grounds.

6. This is not surprising since the human body, and more particularly the human brain, is the most sophisticated set of behaving entities in the known universe.

7. A case in point is Newton's inverse square law of attraction. In *The Analytical Foundations of Celestial Mechanics*, Professor Wintner points out that the simple change from an inverse square law to an inverse cube law introduces a completely new instability parameter into a two-body system (see p. 200). Indeed, with an inverse cube law in the solar system, any planet not moving in a perfect circle around the sun will either spiral inward to collision with the sun or else spiral outward away from the sun (see, for example, H. Pollard, *American Mathematical Monthly*, Vol. 76 (1969), p. 310).

8. When I speak here of 'changes' in a theory, I am not speaking of creative modifications of a theory-in-progress in response to new data or to newly discovered logical relationships, but rather of arbitrary or *ad hoc* tampering with a theory at any given stage of its development.

9. In this regard, G.H. Hardy has remarked: '. . . Greek mathematics is "permanent", more permanent even than Greek literature. Archimedes will be remembered when Aeschylus is forgotten, because languages die and mathematical ideas do not.' (See Hardy's *A Mathematician's Apology*, p. 21). In a similar vein, M.H.A. Newman has said: 'Mathematical language is difficult but imperishable. I do not believe that any Greek scholar of to-day can understand the idiomatic undertones of Plato's dialogues, or the jokes of Aristophanes, as thoroughly as mathematicians can understand every shade of meaning in Archimedes' works.' (See the *Mathematical Gazette*, Vol. 43 (1959), p. 167.)

10. See Carl Friedrich von Weizsäcker, 'Platonic Natural Science in the Course of History', *Main Currents in Modern Thought*, Vol. 29 (1972), pp. 42–52 (translated by Renée Weber).

11. Recent developments in non-equilibrium thermodynamic systems by Professor Prigogine and his collaborators have sometimes been cited as constituting a serious challenge to the physicists' traditional faith in the ultimate simplicity and lawfulness of nature (see, for example, Ilya Prigogine and Isabelle Stengers, *Order Out of Chaos*). In particular, dynamical systems which can be experimentally verified to obey fairly simple mathematical laws (usually expressed by differential equations) when close to equilibrium sometimes exhibit extreme instability and even chaos when they are driven far from equilibrium by externally induced changes in their boundary conditions. In certain cases, such systems reach a bifurcation point beyond which they again exhibit stability (but with significantly increased complexity) and mathematically simple regularities. Because of the empirically observed chaos during the transition to bifurcation, and because there are usually a number of different stable forms which the system can assume after bifurcation, Prigogine argues that there is an essential element of randomness involved in this complexification process.

Of course, as Prigogine implicitly acknowledges, this is basically a philosophical thesis and is not subject to simple experimental verification. It constitutes rather a philosophical interpretation of certain experimental results, an interpretation which can be (and has been) challenged.

A particularly striking challenge to Prigogine's thesis is the recent discovery by a team of French mathematicians working at the University of Strasbourg of surprisingly chaotic behaviour, of a purely mathematical sort, in the evolution of the system of solutions to a fairly simple differential equation, namely, Van der Pol's equation with a parameter a:

$$cx'' + (x^2 - 1)x' + x - a = 0.$$

When the constant c is sufficiently small (but positive), this equation has a limit cycle (a periodic solution) for all values of a between 0 and 1. It has a 'soft' Hopf bifurcation at $a = 1$ (and thus a stable stationary solution when a is greater than or equal to 1). Using methods of nonstandard analysis, the Strasbourg team has shown that, as the parameter a approaches the bifurcation value 1 from below, there appears quite suddenly and violently a radical change in the form and nature of the limit cycle. This novel form of the limit cycle, called a canard (or French duck), is of extremely brief duration and marks a transition from a large cycle (a cycle with relatively large amplitude and period) to a small cycle (with a relatively small amplitude and period). This transition from a large cycle to a small cycle, and the appearance of the canard cycle, occur quite close to the Hopf bifurcation – close enough to be perceived experimentally as part of a 'hard' (catastrophic) bifurcation process. Nevertheless, the canard cycle is not a bifurcation and is mathematically quite distinct from the Hopf bifurcation at $a = 1$.

To get a rough idea of what is involved, let the constant c have the value .05 and suppose the parameter a takes 60 years to cover the distance from 0 to 1. Then the first perceptible variation in the limit cycle will occur about three months before the end, the total evolution from a large cycle to a small cycle with take 20 minutes, and the canard cycle itself will only last for one second (see Callot, Diener, and Diener, 'Chasse au canard', Publications de l'Institut de recherche en mathématique avancée, Strasbourg, 1977, p. 3). Thus, for any empirical system governed by this equation, the canard cycle itself would almost surely remain undetected and would most likely be experienced as a discontinuity in the evolution of the system. Indeed, because of the unusual nature of the canard cycle (it contains both attracting and repelling points of the so-called slow curve associated with the system), the behavior of the system would be experienced as chaotic during the transition from the large cycle to the small cycle. Yet, throughout, the system would, in reality, have been governed by the same global, simple mathematical law.

The importance of this example is that it shows in a particularly clear way how the experimental results of Prigogine can be legitimately regarded as compatible with a non-random interpretation of the evolution of dynamical systems.

12. See in particular Lecture 15, pp. 196–207.

MYTHS, MODELS AND MYSTICISM

1. Some may feel or claim that other species, such as the higher mammals, also have self-awareness. I believe that evidence for this claim is weak. It is mainly the complex, higher-order verbal and symbolic communication between human subjects which allows us to realize that other humans have a subjectivity similar to our own, and animals have so far shown themselves incapable of this kind of communication. Indeed, many researchers feel that self-awarness in humans strictly depends on the individual's capacity for linguistic development. Of course, all of this does not deny the obvious fact that animals do have a form of intelligence and mentation which they demonstrate by such behavior as conditioned learning. But what also seems clear is that human subjectivity is of a nature sufficiently different from whatever animal consciousness may exist to constitute a distinct category and a characteristic feature of the human being.

2. An individual is a subject intrinsically because of the inner world of his subjectivity, but also (partly) extrinsically in relation to *objects*, i.e. that which exists outside of his subjectivity and the subjectivity of others.

3. Early positivism and behaviorism tended to define as real only what existed outside of human imagination and subjectivity, tacitly excluding human consciousness from reality. Such a point of view can easily lead to the absurd position that a human being, with his unique (and, from this point of view, unreal) self-awareness is a stranger, an alien in the universe. Whatever else we may claim to know about the world, we certainly know that it is capable of consciousness because we are part of the universe and we have consciousness. In other words, there can be no consistent, serious doubt about the existence (realness) of subjectivity. The question of the origin and basis of subjectivity is, however, quite another matter, and one about which seriously divergent views can be consistently maintained. Does it arise spontaneously from a sufficiently complex interaction of entities which themselves lack subjectivity, or are there degrees of subjectivity (the *dedans des choses* of Teilhard de Chardin) of which human subjectivity is only the most developed kind accessible to us? The former hypothesis seems rather unlikely and does not provide us with any real explanation of how or at what threshold of complexity self-awareness is born. Nevertheless, it has been defended by some workers in artificial intelligence and some materialistic-minded philosophers who are, perhaps, attracted largely by the apparent possibility of reducing subjectivity to a subcategory of objective reality: the internal states of an individual can be identified, for example, with certain electrochemical configurations within that individual's brain and nervous system which he simply experiences differently than does any outside observer. Such a reductionistic approach to subjectivity seems to miss the point that even if there are observable, physical concomitants to the internal states of an individual, what we actually observe are these physical concomitants and not the self-awareness or subjectivity itself. Yet we each know from our own experience that conscious, internal experience is real. It is the sum total of such conscious, self-aware events that constitutes conscious subjective reality as I have here defined it. However, seeing subjectivity as a universal phenomenon shared in different degrees by all entities is not the only alternative to the materialistic-reductionistic view. Another natural hypothesis is that the locus of human subjectivity is some nonobservable, nonphysical entity, i.e. the soul or spirit of theology and metaphysics, or the self of depth psychology. In any event, resolving the question of the origin and basis of human subjectivity does not seem to be necessary to the development of the central ideas of the present essay.

With regard to my definition of reality: this should not be taken as involving a tacit hypothesis that reality is static or unchanging. To the degree that change and flux exist objectively (rather than merely as subjective perceptions), they are part of objective reality as I have defined it.

4. Of course, strictly speaking we do not have immediate (i.e. 'unmediated') access to the external world. We have immediate access only to the internal states and sensations that our interactions with these external phenomena provoke within us. But we do have direct access to these phenomena in the precise sense that we perceive them directly, i.e. without depending on the reports or beliefs of other subjects like ourselves. In other words, the only subjectivity that directly mediates these phenomena to any given individual is his own.

5. From an epistemological point of view, one of the chief features of this manner of dividing reality is that visible reality is directly accessible to a potentially unlimited number of observers, whereas conscious reality is directly accessible only to one observer. Indirect access is, of course, another matter, which will be the main focus of the present paper. Roughly speaking, the idea is that one can infer the existence of invisible reality from certain behavior of portions of visible reality and that one can infer the existence of unconscious reality from certain individual human behavior together with the individual's verbal reports of his conscious internal states related to and during the given behavior. Notice that the basic division of reality into subjective and objective categories is made from the human point of view, because it is only human subjectivity to which we have even limited access. Thus, any wholly transhuman subjectivity that may exist (e.g. the Mind of God) will, from this point of view, be a part of objective invisible reality.

In thinking about the boundary between visible and invisible reality, we must be careful not to confuse the ultimately unobservable with what may be practically unobservable at the moment. Thus, some remote stars or subatomic particles may be momentarily unobservable but subsequently observable. The point is that we can logically hold that invisible reality exists without believing we can practically determine the boundary between the visible and the invisible at all times. That an ultimately unobservable portion of objective reality exists is therefore a basic philosophical assumption of this essay, but one which, on balance, seems substantially more justified than its negation (i.e. that all objective reality is ultimately observable). See Note 6 as well as the whole discussion of *theories* in the body of the essay. This latter notion allows us to distinguish between raw sense data on the one hand and our (partially subjective) perceptions of such data on the other. See also the discussion of these questions in William Hatcher, 'Science of Religion', *Bahá'í Studies*, vol. 2.

6. All definitions given above are strictly logical in nature and do not themselves involve the assumption that either invisible reality or uncon-

scious reality is nonempty. Indeed, skeptics, positivists, behaviorists, and materialists of various philosophical stripes have often defined their position in part by calling into question the existence of invisible and/or unconscious reality. There are, however, strong arguments against such radical empiricist positions. Though we do not intend to engage in a detailed or systematic discussion of these arguments, some of them can be clearly inferred from the ideas developed and presented in the body of the essay. In any case, the supposition that invisible and unconscious reality do exist underlies the whole of our subsequent discussion, and the skeptical reader can simply take this supposition as a working hypothesis for the remainder of the paper.

7. For example, the invisible force of gravity has significant influence on the observable behavior of free objects in the presence of a large mass such as the earth.

8. What constitutes a 'reasonably accurate' picture of reality will often depend considerably on pragmatic considerations, i.e. on what our needs are at the given time and in the given circumstances. A particular mental model may be sufficiently accurate to allow for correct predictions within certain limits of tolerance but insufficient if pushed beyond these limits. For example, a rather crude model of the invisible force of gravity may be sufficient to prevent us from deliberately walking off cliffs and may even allow us to predict with some accuracy the trajectory of thrown objects. But it might not allow us to explain the movement of a pendulum or the motion of the planets. The point is that we can consistently suppose that reality has a definite structure (philosophical realism) without having to believe that we will necessarily ever arrive at a perfectly accurate understanding of that structure. Some philosophers (e.g. Hilary Putnam) have tried to argue that realism implies belief in the existence of a 'perfect correspondence' between reality and the human mind and have then cited the (rather strong) evidence against the possibility of such a perfect correspondence as evidence against realism. (See, for example, Putnam's 'Realism and Reason', *Proceedings and Addresses of the American Philosophical Association* 50.6 (1977):483–98.) G.H. Merrill has given a very cogent refutation of Putnam's main arguments by showing that the existence of mind-independent unobservable structure and the possibility of our misconception of that structure, are both consistent with the usual model-theoretic notion of truth (see G.H. Merrill, 'The Model-Theoretic Argument against Realism', *Philosophy of Science* 47 (1980):69–81). To Merrill's discussion, I would add the further observation that the objective structure of reality may well be much more subtle than Merrill supposes in his article. It may, for example, involve a (possibly unbounded number of) infinitary relations and operations, as well as finitary ones. This would mean that, in a finitary language, we could never talk about more than a

small portion of reality at any given time (i.e. under any given interpretation of the language). Of course, this observation is not a criticism of Merrill. On the contrary, his refutation of Putnam, supposing a structure involving only finitary relations, shows that appeal to infinitary relations and operations is not logically necessary to justify realism.

9. A theory is usually *presented* as a set of propositions that make affirmations about reality. If the theory is true, then the propositions which comprise the *presentation* will, in general, contain both *concrete* terms, i.e., terms that designate observable entities or forces, and *abstract* (also called *theoretical*) terms, i.e. terms that designate invisible or subjective configurations. The extreme form of philosophical materialism which holds invisible reality to be nonexistent is often formulated by the affirmation that all abstract terms can, in principle, be eliminated from scientific discourse. There is, at present, considerable evidence that this is not possible and that abstract terms are unavoidable in the presentation of adequate theories of physical reality. This and related points will be discussed later in the present essay.

It should also be borne in mind that, just as the boundary between the observable and the nonobservable is often unclear, so the distinctions between the abstract and the concrete, the theoretical and the factual, are likewise relative rather than absolute. The problem is that by the time an individual has developed the social and linguistic skills necessary to communicate personal observations to others, she or he has undergone such a degree of socialization that it is impossible to distinguish the purely observable raw sense data from the theoretical framework which allows (and, indeed, constrains) the individual to perceive these data in a particular way. We are left only with certain statistical correlations between the reports of different observers and are thereby forced to engage in further theorizing in order to distill some generally accepted 'objective' content common to most observations by different individual subjects.

10. Indeed, it follows from results of modern logic that there is usually an infinite number of logically incompatible theories consistent with any finite set of facts (i.e. observational statements). Yet conceiving of even one plausible theory can be very difficult. Thus, no (finite) collection of observational statements determines a unique theory of invisible reality, and there is no formula for constructing theories from facts. The leap from fact to theory is a leap of the imagination. Theory making is therefore one of the most creative of all human endeavors.

11. It should be quite clear how my definition of the notion of a myth differs from the popular conception of a myth as a false, fanciful, or even absurd theory. Defenders of the social value of myths point out that a myth may express important social or psychological truths even when false under its literal interpretation. In particular, the claim has been made with

increasing frequency in recent literature on the subject that modern sociologists and anthropologists may have judged many scientifically primitive, mythmaking societies unfairly by imposing literal interpretations on myths that should rather be understood in a more allegorical or metaphorical fashion as expressing truths about subjective reality. On the basis of this perceived injustice, some have criticized my use of the term saying, in effect, that virtually all myths are true when properly interpreted and that myths are viewed as false only by those who try to understand them from a fundamentally prejudiced, unsympathic, or overly materialistic point of view.

While I am quite sympathetic to the spirit of this criticism, I believe it is based on several misconceptions. In the first place, it is largely directed against the popular conception of myths as false theories, and I cannot help but feel that those who have criticized me in this regard have not really assimilated the difference between my definition and the commonly received one. Moreover, since a myth, in my sense of the term, is a theory that society accepts because it perceives the theory as describing a need-satisfying configuration, my approach provides a reasonable framework for understanding how subjective input enters the mythmaking process. In any case, all these issues must not be allowed to obscure the even more basic point that the same theory may be true under one interpretation and false under another. Thus, a myth may well be false of the phenomenon it ostensibly purports to describe but true when interpreted so as to apply to another (possibly subjective) phenomenon. However, to insist that a theory (and in particular a myth) must be regarded as valid if true under *some* interpretation is to deny ourselves the vocabulary necessary to an adequate and clear discourse on the whole question.

12. The notion of a sterile theory will be raised again and can be better understood in the context of the discussion of truth criteria on p. 32ff.

13. In other words, the scientific revolution was, fundamentally and essentially, a social revolution because it was based on revolutionary changes in the way (some) societies behaved in certain important respects. Of course, everyone admits that modern science has wrought significant social changes, but the tendency has been to attribute these changes either to technology (i.e. to the material changes resulting from the practical applications of scientific knowledge) or else to the so-called mechanistic worldview that grew out of the successes of sixteenth and seventeenth-century physics and was the philosophical precursor of positivism and other modern forms of philosophical materialism. My identification of science with the enterprise of model building is different from these more traditional viewpoints primarily because it sees successful science as a *result* of social change (i.e. the replacement of one set of social behaviors and attitudes by another) rather than only as a cause of social change.

My identification of science with the enterprise of model building will probably be unacceptable to any scientist who believes that science is inherently antimetaphysical in one way or another. It is precisely my intention to show that science, rightly conceived, is not intrinsically materialistic or reductionistic, and that the model-building paradigm provides a coherent framework in which to interpret science independently of any pro- or anti-metaphysical thesis. This point should become clearer as our discussion proceeds.

Some readers of preliminary versions of this esasy have felt that I attribute undue importance to the transition from mythmaking to model building. For example, some feel that early societies show a much greater degree of scientific thinking than I appear to acknowledge, while others feel that many aspects of modern scientific practice are uncomfortably close to what I have called mythmaking. I believe these perceptions reflect primarily a basic difference in historical viewpoint between these readers and myself, and that is one of the reasons (though not the only or even the most important reason) why this essay culminates in a discussion of what I call the organismic view of history, which sees human history as a sequence of 'growth stages' analogous to similar stages in the life of an individual.

Clearly there are examples of scientific thinking all through history, and it would be just as absurd to imagine that a social revolution had no precursors as it would to expect a human infant to remain unchanged for the first twelve years of life and then to become a fully developed adolescent overnight. But, continuing the analogy, just as an adolescent develops capacities and powers which are incomparably greater than their pre-adolescent counterparts, so (I believe at any rate) the persistent, systematic, and socially generalized applications of model-building processes that characterize the last four hundred years (and especially the last one hundred and fifty years) are significantly greater than any of their historical antecedents.

Concerning the question of similarities between the current behavior of practicing scientists and that of prior or contemporary non-scientists, we should recall that both myths and models are first of all theories and thus naturally elicit many similar human responses. My point is that, in spite of these pervasive similarities, there is at least one fundamental way in which building models differs from elaborating myths (i.e. in giving priority to truth instead of attractiveness when processing new theories) and that this difference is sufficient to explain the impressive success of modern science.

14. See Note 9 above.

15. Idealization and interpretation are thus inverse to each other. Idealization moves 'inward' from the perception of the phenomenon to the formation and conception of the theory within the confines of human subjectivity. Interpretation moves 'outward' from the theory to the phenomenon.

16. This is not to say that emotional attachment to a theory is

inherently incompatible with model building. Indeed, it is only natural and healthy that we should cherish a theory which we have lovingly and carefully developed and which has served us well in our ongoing efforts to increase our knowledge of reality. But our strongest emotional attachment should be to the model-building process itself (and to the truth it brings) rather than to any particular theory. We must, therefore, always be alert to the possibility of our love for a particular theory being transmuted into a truth-corrupting idolatry.

17. See Karl Peters, 'Religion and an Evolutionary Theory of Knowledge', *Zygon* 17.4(1982):385–415.

18. The section below on mysticism discusses in more detail how one can approach the question of transcendent experience within the framework developed in this essay.

19. See, below, pp. 94–122: 'Science and the Bahá'í Faith'.

20. The basic source for the theory of progressive revelation is Bahá'u'lláh, *Kitáb-i-Íqán: The Book of Certitude*. My own understanding of Bahá'u'lláh's ideas has been influenced in various degrees by the thinking of Arnold Toynbee (see in particular his *Christianity among the Religions of the World*) and Henri Bergson, especially his *The Two Sources of Morality and Religion*.

21. Although it seems clear that some forms of animism represent what I have called 'common-denominator religions' or taboo systems, I am not at all sure that every form of animism falls into this category, and there may well be genuine revelatory elements in many so-called primitive religions. No doubt, there has been much arrogance and prejudice involved in the study of these religious phenomena by social scientists, and we should, I feel, be quite prudent in making judgements about them.

22. These embellishments may vary quite a bit in detail, according to the various historical circumstances involved in each particular case, but they usually have the effect of glorifying the community of believers, in one way or another, as a people superior to all non-believers.

23. *Paris Talks*, p. 144.

24. ibid. pp. 143–4.

25. *The Promulgation of Universal Peace*, p. 181.

26. *Some Answered Questions*, pp. 251–2.

27. See, for example, Fritjof Capra, *The Turning Point*, and Paul Davies, *God and the New Physics*.

28. Of course, experience is part of knowledge, and certainly valid experience of invisible reality would help to construct accurate models of invisible reality. Thus, mysticism may be properly thought of as part of religion. Nevertheless, it is logically possible that we might be successful in building an accurate model of some portion of invisible reality without ever experiencing directly that reality. In other words, success in the religious enterprise neither logically implies nor logically depends on

success in the mystic enterprise. We thus avoid the frequently made identification of religion with mysticism.

29. Widespread similarities among mystic experiences, even if shown to exist, would not necessarily imply the existence of some common extrinsic origin of these experiences. For example, such similarities could be accounted for by regularities in human psychology and brain physiology. What I am suggesting here is rather a negative test, namely, that widespread and radical dissimilarities in certain mystical experiences might reasonably be taken as evidence against the existence of a common extrinsic origin to those experiences.

30. See, for example, Rudy Rucker, *Infinity and the Mind* and Fritjof Capra, *The Tao of Physics*.

31. Not only scientists but some religious thinkers as well have suggested that extreme caution should be used in gauging the validity of one's inner experiences. For example, the Bahá'í authority and author Shoghi Effendi has said, in answer to an inquiry concerning the validity of various types of inner experience:

You yourself must surely know that modern psychology has taught that the capacity of the human mind for believing what it imagines, is almost infinite. Because people think they have a certain type of experience, think they remember something of a previous life, does not mean they actually had the experience or existed previously. The power of their mind would be quite sufficient to make them believe such a thing had happened. (Quoted in Hatcher, *The Concept of Spirituality: Bahá'í Studies*, vol. 11, Note 35.)

There are other statements by Shoghi Effendi in a similar vein. Such statements are made all the more significant by the fact that Shoghi Effendi elsewhere stresses the legitimacy, and indeed the fundamental and vital character, of mystic experience as an essential part of religion: '. . . the core of religious faith is that mystic feeling which unites Man with God . . . The Bahá'í Faith, like all other Divine Religions, is . . . fundamentally mystic in character.' (*Directives from the Guardian*, pp. 86–7).

32. My basic source here is Shoghi Effendi, *The World Order of Bahá'u'lláh*, as well as the authors and references in Note 20 above. However, some of the concepts used in the ensuing discussion of the organismic theory of history are (as far as I know) my own.

33. For some of these ideas, see R. Leakey and R. Lewin, *People of the Lake*.

34. Shoghi Effendi, *The World Order of Bahá'u'lláh*, p. 202.

35. ibid. pp. 163–4.

36. A 1985 statement, *The Promise of World Peace*, by the Universal House of Justice (the supreme legislative body of the Bahá'í Faith) contains a more complete discussion of these important questions.

37. 'Abdu'l-Bahá, *The Secret of Divine Civilization*, pp. 66–7.

FROM METAPHYSICS TO LOGIC

1. A cosmological proof is one which, in at least one of its assumptions, invokes empirical data about physical reality (i.e. the cosmos). Frequently, a proof is called 'cosmological' to distinguish it from an 'ontological' proof, where this latter means a proof involving only abstract logical or metaphysical principles whose validity is regarded as *a priori*, self-evident and independent of the material world.

2. See Aristotle, *Works*, Vol. II, *Physica*, 258b10 ff., and Vol. VIII, *Metaphysica*, 994a1ff.

3. It is only fair to stress that Aristotle was not dealing with an arbitrary infinite regression, but an infinite regression of causes. Moreover, Aristotle's philosophy contained an exhaustive analysis of the nature of causation, and it is on the basis of this analysis that he declares an infinite regression of causes to be impossible.

4. For instance, systems of interacting elementary physical particles exhibit highly complex mutual causation that sometimes appears incompatible with a strictly linear notion of causality.

5. See H.A. Davidson, 'Avicenna's Proof of the Existence of God as a Necessarily Existent Being', *Islamic Philosophical Theology*, ed. P. Morewedge, SUNY Press, Albany, N.Y., 1979, 165–187, p. 171. Our exposition of Avicenna's proof in the present essay relies heavily on Davidson's excellent article, which is based on Avicenna's *Najat (Salvation)*, Cairo, 1938, pp. 224ff. and his *Kitab al-Isharat wa-l-Tanbihat (Book of Counsels and Commentaries)*, ed. J. Forget, Leiden, 1892, pp. 140ff. We also make significant use of the French translation of this latter work by A. Goichon, *Livre des directives et remarques*, Beirut-Paris, 1951, pp. 51ff., as well as Avicenna's *La métaphysique du Shifa*, trans. M. Anawati, Institut d'études médiévales, Université de Montréal, 1952, *Livre huitième*. However, in line with an accurate observation of Davidson (op. cit., p. 175), to the effect that the concept of necessary existence is superfluous to the logic of Avicenna's proof, we have simplified our exposition by entirely omitting reference to necessary existence, retaining only the distinctions between caused and uncaused, simple and composite entities. Thus, our exposition and subsequent analysis of Avicenna's proof are based strictly on classical, truth-functional logic, avoiding completely the various controversies concerning the modal logics traditionally used in metaphysical proofs of God's existence.

6. However, see Aristotle, *Metaphysica*, op. cit. 1072a20 ff.

7. See Davidson, op. cit. p. 175.

8. Ibid.

9. Ibid. pp. 175–6.

10. Ibid. p. 176.

11. Because physical entities are composite, and no composite entity can be uncaused, as explained above.

12. Because there is only one uncaused cause, because of the unique properties it has (as described above), and because (as will be presently shown) it is the ultimate cause of every other entity in existence.

13. Davidson, op. cit. pp. 178–80. Clearly, E is the only entity that exists outside of C. Since E is the cause of the collection C, it is the (indirect) cause of every member of C. Furthermore, E is its own cause. It is therefore the ultimate cause of every entity in existence (see also ibid. p. 177).

14. Davidson also identifies this strategy of Avicenna as the crucial step in Avicenna's proof, noting that '. . . the cogency of his [Avicenna's] argument depends upon the legitimacy of that procedure'. (ibid., p. 179). However, Davidson does not say whether or not he regards 'that procedure' to be legitimate.

15. Beginning in the 1870s, Cantor developed an abstract theory of *sets* (arbitrary collections of arbitrary objects) which eventually had enormous impact on both logic and mathematics. It also raised a host of new and difficult philosophical, logical, and mathematical questions, some of which are implicitly raised by the method of Avicenna's proof.

16. This point may appear at first to be logical hairsplitting, but it is not. We infer from our experience and observation of various processes in the material world that physical phenomena are connected with each other by means of highly complex causality relationships. Thus, when observing some phenomenon accessible to us in our local space-time framework, we often ask the question 'what is the cause of this phenomenon?'. In asking this question, we usually presume (rightly or wrongly) that the cause will be some other phenomenon (within the same interacting system that constitutes our universe). But, if we ask the question 'what is the cause of the whole system (the universe) itself?' we have jumped to another level, for now either some part of the universe is the cause of the whole (including the part in question) or else there is something outside the universe that is the universe's cause. (But how can something exist *outside of the universe* if by 'the universe' we mean everything in existence, material or not?) We have made a transition from 'local causality *within* a system' to 'global causality *of* the entire system itself'. Essentially the same point was made by Bertrand Russell in a famous debate with F.C. Copleston, 'The Existence of God', on the Third Programme of the B.B.C. in 1948 (see B. Russell, *Why I am Not a Christian*, 144–168, particularly pages 151–5). However, I do not agree with Russell's contention that the question of the existence of a cause of the universe is meaningless. (It is worthwhile to note that, in this debate, Copleston uses a variant, due to de Leibniz, of Avicenna's proof of God's existence.)

17. As it turns out, there are non-contradictory, formal systems of set theory in which self-membership is possible. However, nobody has any idea of how to meaningfully interpret these paradoxical systems as representing collections of real objects. (Cf. the interesting discussion of this point by L. S. Moss in the *Bulletin of the American Mathematical Society*, Vol. 20, No. 2 (1989), pp. 216–225.)

18. According to most notions of entityship, the class C of all caused entities other than C would not be considered an entity, for it does not seem to possess that internal cohesion we usually associate with whole objects. But it is even more difficult to conceive of a notion of entityship that would accept C but reject C^* and C^{**}. Notice that we have not bothered to consider the entityship question for the class C^+ obtained by adding just E to C.

19. We will attempt to assess the reasonableness of the contingency principle later on. We have already encountered it in the course of our discussion of Avicenna's proof, which assumed the following stronger *composite causation principle*: No composite phenomenon can be uncaused. We avoid assuming this latter principle because our logical analysis of causation will allow us to deduce it logically from the contingency principle (see Lemma 1 in the following).

20. With regard to the discussion on global causality in Note 16 above, we have now chosen the first of the two alternatives mentioned; that is, we use the term 'universe' to refer to the collection of all existing entities. Thus, God, if He exists, is part of the universe.

21. Except for the terms 'composite', 'component', 'simple,' and 'phenomenon', which are taken from metaphysics, all terms defined in this paragraph are taken from set theory. The distinction between 'class' and 'set' was originated by Cantor, and was developed and elaborated by Von Neumann, Gödel, and Bernays. The notion of 'individual' originates, as far as I know, with Bertrand Russell and was subsequently developed by E. Zermelo, who called them urelements. For discussion of these distinctions as they are currently used in mathematics and logic, see W.S. Hatcher, *The Logical Foundations of Mathematics*, Pergamon, Oxford, 1982. A good working model for our ontology is to let the class V be the collection of all hereditarily finite sets in any model of Zermelo-Fraenkel set theory with at least one urelement. However, it is important to realize that, in defining these various ontological categories, we are not positing their existence. In particular, we have nowhere explicitly assumed (nor will we do so) that any simple entity exists. Rather, we will prove the existence of (at least one) simple entity on the basis of our subsequently-defined causation principles.

22. For a more detailed discussion of well-foundedness in set theory see the work cited in Note 21 above. However, all we will really need is the principle that no class (composite) can be a member (component) of itself,

and the reader can just take this as the working definition of well-foundedness for the purposes of this article. The problem of self-membership and the justification for excluding it have already been discussed in the preceding section on evaluating Avicenna's proof.

23. The intuitive notion underlying these principles is that any cause A of a caused phenomenon B must, in some sense, be greater (stronger, more potent) than B, for otherwise B would be sufficient to itself, i.e. uncaused. The principles of causality and transitivity are logically equivalent to the following more complicated (but perhaps more intuitively evident) principles. (1) quasi-irreflexivity: A causes A if and only if A has no other cause B; (2) antisymmetry: If A causes B and A $\neq$ B, then B does not cause A (or, equivalently, A $\rightarrow$ B $\rightarrow$ A implies A = B); (3) strict transitivity: If A, B, and C are all distinct and A $\rightarrow$ B $\rightarrow$ C, then A $\rightarrow$ C.

24. The intuitive justification of the potency principle is that any phenomenon A which is sufficiently potent to cause a phenomenon B must, *a fortiori*, be strong enough to cause any portion (subphenomenon) or any component of B. It is important to note that the partial converse of the potency principle which asserts that a cause A of every component of B is a cause of B is false. In other words, to be a cause of a composite phenomenon B, it is not sufficient to be a cause of every component of B; the whole (i.e. B) is greater than the sum of its parts (components). Of course, the (total) converse of the potency principle is (trivially) true since every phenomenon B is a sub-phenomenon of itself. Notice that, in general, it will not be the case that a composite phenomenon is the cause of its subphenomena or even of its components.

25. The search, in modern physics, for a unified field is thus somewhat analogous to our search for God, because a unified field is defined as a single force from which each of the four fundamental forces can be derived. However, we do not presume that the force-entity God is a physical force. Moreover, it is by no means certain that the four fundamental forces of modern-day physics are all of the physical forces that will ever be discovered to exist. (Nor is it clear that the functioning of higher-order phenomena, such as complex living organisms or the human brain, can really be explained as resulting from some combination of the four forces of physics.) But, the most fundamental difference of all is that God is defined as the global cause of the universe itself (as a whole), and not just of the entities within the universe (see the discussion of this point in note 24 above).

26. Notice, however, that the existence of a universal uncaused cause A is incompatible with the existence of any non-universal uncaused cause B. Indeed, to be universal, A must be a cause of every phenomenon, including B. But B is uncaused and, by the causality principle, cannot be other-caused. Thus, Aristotle's uncaused cause is the only candidate for Godhood.

This observation shows that Aristotle's identification of his prime mover with God was much less gratuitous than might appear at first.

27. Because the following argument is a bit more complicated than is usual for philosophical discussions, we temporarily adopt a more formal, mathematical style of exposition.

28. Notice that this is the only instance in our entire argument where we appeal to the transitivity principle, and even this instance can be avoided as follows: Having established that $B \rightarrow A \rightarrow C$, where B is an uncaused, simple entity, we form the composite phenomenon C^* by adjoining the entity B to the collection C of all caused entities. Then, by Lemma 2, $B \rightarrow C^*$ and, by the potency principle, $B \rightarrow C$. Thus, the transitivity principle is not logically necessary to our argument, and we have assumed it primarily because it is so natural and seems to facilitate our intuitive grasp of the causality relation. That it can be avoided is significant however, because it shows that the existence of a universal uncaused cause is not logically dependent on a strictly linear notion of causality (cf. Note 4 above).

29. The contingency principle is a stronger assumption than the others because it is inductive — generalizing from particular entities to a whole class of entities — rather than logical or analytic (as is the potency principle, which goes from an established whole to its parts).

30. The hypothesis that the physical universe has always existed is logically compatible with either a caused or an uncaused universe. Indeed, the idea that the physical universe is caused but eternally existing is one of the teachings of the Bahá'í Faith (see 'Abdu'l-Bahá, *Some Answered Questions*, pp. 180ff.). However, an uncaused universe does not seem compatible with the notion that the universe had a discrete beginning in time.

31. Based on a penetrating and illuminating discussion of 'Abdu'l-Bahá (see Bahá'u'lláh and 'Abdu'l-Bahá, *Bahá'í World Faith*. pp. 336ff.), we have given elsewhere a more careful and detailed proof of God's existence, using such scientific principles as the second law of thermodynamics. (See W.S. Hatcher, 'The Unity of Religion and Science', in the monograph *The Science of Religion*, 15–28, reprinted with minor revisions in *The Bahá'í World*, vol. 17, Bahá'í World Centre, Haifa, 1981, 607–619.)

A LOGICAL SOLUTION TO THE PROBLEM OF EVIL

1. In this article I will use the following signs for the sentential connectives: $\supset$ for 'if . . . then . . .'; $\wedge$ for 'and'; $\vee$ for 'or'; $-$ for 'not'; $\equiv$ for 'if and only if.' Read the existential quantifier (Ey) as 'there is at least one y such that' and the universal quantifier (x) as 'for all x' or 'no matter what x we choose.'

2. $(E!x)$ is read as 'there exists one and only one x such that,' and

'$\iota x F(x)$' is read as 'the unique thing x such that $F(x)$ is true.'

3. See 'Abdu'l-Bahá, *Some Answered Questions*, pp. 214–16, 263–4.

SCIENCE AND THE BAHÁ'Í FAITH

1. Shoghi Effendi, *World Order of Bahá'u'lláh*, p. xi; italics added.

2. Bahá'u'lláh, *Gleanings from the Writings of Bahá'u'lláh*, pp. 194–95.

3. Bahá'u'lláh and 'Abdu'l-Bahá, *Bahá'í World Faith*, pp. 382–83.

4. 'Abdu'l-Bahá, *Some Answered Questions*, pp. 158–9.

5. This is a conscious paraphrase of a description due to W. V. Quine, *Word and Object*, p. 3.

6. For a much more detailed and exhaustive analysis of this conception of scientific method see my 'Science and Religion', *World Order* 3 (Spring 1969): 7–19 (reprinted in *The Science of Religion: Bahá'í Studies* 2, 1–13).

7. Some might feel that deductive logical proofs are absolute, but such proofs proceed from premises which are based ultimately on empirical and thus inductive or probable inference. See ibid. for a more detailed analysis and discussion of these points.

8. The appeal to probable inference here is in the sense of 'approximate' and not in the technical sense of the strict construction of a probabilistic model for the phenomenon being investigated. Probability in our sense is thus a measure of the relative ignorance of the knowing subject rather than the hypothesis that the phenomenon under investigation is indeterminate in some way. This leaves unanswered the question of whether every use of probability can be so regarded. However, if one espouses an essentially pragmatic epistemology, as I do, it may not even be necessary to determine, in any given instance, which part of our world view comes from the viewer and which part derives from the thing viewed. We have only to evaluate the explanatory and predictive value of our model according to pragmatic criteria. (See my 'Foundations as a Branch of Mathematics', *Journal of Philosophical Logic* 1 (1972): 349–58, for a further discussion of these points. Cf. also the discussion in 'Platonism and Pragmatism' above.)

9. 'Abdu'l-Bahá, *The Promulgation of Universal Peace*, p. 253.

10. It is interesting to note the discussion given of the use of scriptural authority. In *Some Answered Questions*, pp. 298–299, 'Abdu'l-Bahá points out that man's understanding of scripture is limited by his own powers of reasoning and interpretation. Since these powers are relative, so is his understanding of scripture. Thus, regardless of the authority one attributes to the text itself, arguments based on such authority are in reality based on man's understanding of the text and hence are not absolute.

11. 'Abdu'l-Bahá, *The Promulgation of Universal Peace*, p. 255.

12. 'Abdu'l-Bahá, *Some Answered Questions*, p. 157.

13. ibid. pp. 157–8.

14. ibid. pp. 220–221.

15. We have in effect a Platonic metaphysics combined with a pragmatic epistemology, the essential connection between the two being the Manifestation. See also Note 30 below.

16. Of course it is clear that such things as remote stars and subatomic particles are not immediately accessible, but the refined techniques used to study them are often appealed to as concrete extensions of the immediately accessible, even to the extent of identifying the object of study as being the techniques themselves (operationalism). On the other hand such examples (and especially the subatomic case) can be seen already as a partial refutation of the narrow view of scientific method. Witness the difficulty encountered by positivistic philosophers of science in assimilating the study of these phenomena to the narrow view.

17. The most well-known attempts are those of the Vienna-Oxford school typified in Alfred J. Ayer's *Language, Truth, and Logic*.

18. Comparison may well be made here between such an experience and that of mystics. Perhaps the mystic is initially overwhelmed by the newness and intensity of his first experience and thus is led to feel that it is essentially and irredeemably chaotic and unsystematic. This would naturally lead to the glorification of the subjective which is characteristic of the existentialist view as well as to the conviction that mystic experience is essentially nonobjectifiable. But it is precisely my suggestion that the building of a religious community of understanding in a scientific way can lead to a relative objectification of mystic experience similar to that effected by the application of scientific method to other levels of experience. The resulting framework of interpretation would allow the individual to proceed from the initial mystic experience to a new stage of spiritual perception or knowledge, again bringing order out of chaos. This model also serves to illumine the relationship between the individual practicant and the community. The individual's mystic experience is his own and no one else's, but he has to relate properly to the community if his internal experience is to be of genuine profit to him. At the same time there is the further benefit to the community itself, which profits from harnessing the individual's spirituality in the form of service.

19. One thousand years is mentioned in the Bahá'í writings as representing an approximate length of time between two successive occurrences of revelation within a given collective or social gestalt. However, it is stated that this is an approximate or average time span which can vary and which in fact has varied in history. Also, as the collective awareness of human society has increased through progressively more sophisticated means of transportation and communication, traditional gestalts widen, overlap, and fuse, lessening thereby the necessity for parallel or complementary occurrences of revelation.

20. In this regard Bahá'u'lláh has given the following clear statement: 'Beware, O believers in the Unity of God, lest ye be tempted to make any distinction between any of the Manifestations of His Cause, or to discriminate against the signs that have accompanied and proclaimed their Revelation. This indeed is the true meaning of Divine Unity, if ye be of them that apprehend and believe this truth. Be ye assured, moreover, that the works and acts of each and every one of these Manifestations of God, nay whatever pertaineth unto them, and whatsoever they may manifest in the future, are all ordained by God, and are a reflection of His Will and Purpose. Whoso maketh the slightest possible difference between their persons, their words, their messages, their acts and manners, hath indeed disbelieved in God, hath repudiated His signs, and betrayed the Cause of His Messengers.' (*Gleanings*, p. 59.)

21. My brief discussion of the Bahá'í concept of progressive revelation does not address itself directly to a number of questions which a thoughtful reader may be naturally led to pose. To treat these questions within the confines of a short paper like this would be impossible, and such excursions also would blur the sharp forcus that is the proper goal of any essay. One important question, which is only partially treated in the foregoing, is that of establishing criteria for recognizing valid occurrences of the phenomenon of revelation. It is interesting to note that this and other related questions are treated in considerable detail in the writings of the Báb, Bahá'u'lláh, and 'Abdu'l-Bahá to which the reader is referred. Although these writers make some references to the internal states of the Manifestations, the criteria they give for assessing any claim to revelation mostly involve observable events. Besides the person of the Manifestation, his life, his teachings, his influence, and the social organization and civilization based upon them, one of the most important characteristics which these writers associate with authentic revelation is the Manifestation's capacity for 'revealed writing'. This latter refers to the manner of writing (spontaneous and uninterrupted), the quantity and volume of writing, the capacity to reveal writing under all conditions of human life and without the benefit of formal schooling, and, most important, the spiritual and literary quality, the depth, the cogency, and the rationality of the content of the writing. Thus, e.g. Bahá'u'lláh left well over one hundred major works of writings, some of them written while in prison, in chains, or under other extreme conditions. Moreover, he had no formal schooling whatever beyond learning to read and write his native language of Persian. One of his major works, the *Book of Certitude*, whose English translation runs to over two hundred pages, was written in the space of two days and two nights. Since these writings are published in many languages and widely disseminated, there is a maximum opportunity for objective verification of their quality and depth. The original manuscripts are all preserved, and there is consequently no question of interpolation or of other modifications done

before publication. For an excellent discussion of these and other related points, together with eyewitness accounts and photocopies of many archival materials, see A. Taherzadeh, *The Revelation of Bahá'u'lláh*, 4 vols. (Oxford: George Ronald, 1974–88). Another important point stressed by Bahá'u'lláh and 'Abdu'l-Bahá is that a Manifestation is the first to practice his own teachings. He is the first example who lives his teachings into reality, whereas many philosophers, scientists, thinkers and creative artists produce their works while living lives widely at variance with the precepts or ideals these works seek to express. In particular the Bahá'í concept of revelation must not be confused with a host of other phenomena which are sometimes popularly called 'revelation'. I am thinking of such things as trances, occultism, hypnotism, various psychopathological states, etc. As I have tried to make clear in my discussion, 'revelation' in the Bahá'í concept refers to a naturally occurring periodic phenomenon (of rather long period) and not to abnormal or occult events. Of course the laws governing occurrences of revelation are viewed by Bahá'ís as depending on the will of God, but this is no less the case for all natural laws, and so revelation would have no special status in this regard. I feel that these supplementary comments are made necessary primarily because of the current resurgence of occultism, witchcraft, satanism, and other such activities which are specifically condemned by Bahá'u'lláh and 'Abdu'l-Bahá as superstitious and based on false imagination. Such popular fascination with the 'supernormal' tends to create an ethos in which objective discussion of questions relating to religious experience becomes difficult and the otherwise clear lines between authentic spirituality and superstitious exoticism obscured.

22. The revelation of Jesus was focused primarily on the individual and can be viewed at least in part as a counterbalance to the overemphasis on the totalitarian state and to the miserable social conditions and status to which the majority of the recipients of his message were subject.

23. Bahá'u'lláh does not claim to be the last of these messengers, for according to his teachings the succession will never stop; nor will human and social evolution ever come to a dead end (though the ultimate physical death of the solar system itself seems inevitable according to the best current scientific knowledge). However, he does state clearly that the next Manifestation will not come before the lapse of a thousand years' time.

24. This reflects a fundamental principle of evolutionary phenomena: That which is functional and productive at one stage of the process can become dysfunctional and unproductive at another stage. The same principle can be applied in attempting to understand the various changes in religious practice wrought by each successive revelation.

25. With regard to the individual purpose of religion Bahá'u'lláh has said: 'Through the Teachings of this Day Star of Truth [the Manifestation]

every man will advance and develop until he attaineth the station at which he can manifest all the potential forces with which his inmost true self hath been endowed. It is for this very purpose that in every age and dispensation the Prophets of God and His chosen Ones have appeared amongst men . . .' (Bahá'u'lláh, *Gleanings*, p. 68.)

26. Bahá'u'lláh and 'Abdu'l-Bahá, *Divine Art of Living*, rev. ed., p. 92.

27. Nothing that I have said in the foregoing should be taken as implying that the aesthetic and emotional aspects of religion should in any way be deemphasized, neglected, or excised from religion. My contention rather has been that when religion is excluded from the application of scientific method the aesthetic and emotional tend to become drastically overemphasized as they are then seen as constituting the only datum of religion. But it is my feeling that when a more balanced picture of religion is attained and its basically cognitive nature is recognized then these other aspects naturally fall into place in a healthy way, neither being indulged or sought for their own sake on the one hand nor rejected on the other. I think it is fair to say that many of the excesses witnessed throughout religious history, such as fanaticism, asceticism, mystic thrill seeking, and withdrawal from society, can be attributed largely to the lack of the sort of balanced viewpoint I am seeking to describe. It is interesting to note that Bahá'u'lláh pointedly condemns these specific excesses as well as others.

28. Bahá'u'lláh, *Gleanings*, pp. 164–66.

29. In this connection, Bahá'u'lláh has said: '. . . all things, in their inmost reality, testify to the revelation of the names and attributes of God within them . . . Man, the noblest and most perfect of all created things, excelleth them all in the intensity of this revelation, and is a fuller expression of its glory. And of all men, the most accomplished, the most distinguished, and the most excellent are the Manifestations of the Sun of Truth. Nay, all else besides these Manifestations, live by the operation of their Will, and move and have their being through the outpourings of their grace: (ibid. pp. 178–79.)

30. The crucial role of the Manifestation as the link between the transcendent absolute reality and the world of man is expressed by 'Abdu'l-Bahá: 'The knowledge of the Reality of the Divinity is impossible and unattainable, but the knowledge of the Manifestations of God is the knowledge of God, for the bounties, splendors, and divine attributes are apparent in them. Therefore, if man attains to the knowledge of the Manifestations of God, he will attain to the knowledge of God; and if he be neglectful of the knowledge of the Holy Manifestation, he will be bereft of the knowledge of God'. ('Abdu'l-Bahá, *Some Answered Questions*, p. 222.)

31. 'Abdu'l-Bahá, *Paris Talks*, pp. 143–46.